S0-GYC-235

A typology of sculpted forms in open bedrock channels

by

Keith Richardson
Lancaster Environment Center
Lancaster University
Lancaster, LA1 4YB
UK

Paul Anthony Carling
School of Geography
University of Southampton
Highfield, Southampton, SO17 1BJ
UK

THE GEOLOGICAL SOCIETY OF AMERICA

GB
561
R53
2005

Special Paper 392

3300 Penrose Place, P.O. Box 9140 ▪ Boulder, Colorado 80301-9140, USA

2005

OHIO UNIVERSITY LIBRARY
WITHDRAWN
3C 9/28/05

Copyright © 2005, The Geological Society of America, Inc. (GSA). All rights reserved. GSA grants permission to individual scientists to make unlimited photocopies of one or more items from this volume for noncommercial purposes advancing science or education, including classroom use. For permission to make photocopies of any item in this volume for other noncommercial, nonprofit purposes, contact the Geological Society of America. Written permission is required from GSA for all other forms of capture or reproduction of any item in the volume including, but not limited to, all types of electronic or digital scanning or other digital or manual transformation of articles or any portion thereof, such as abstracts, into computer-readable and/or transmittable form for personal or corporate use, either noncommercial or commercial, for-profit or otherwise. Send permission requests to GSA Copyright Permissions, 3300 Penrose Place, P.O. Box 9140, Boulder, Colorado 80301-9140, USA.

Copyright is not claimed on any material prepared wholly by government employees within the scope of their employment.

Published by The Geological Society of America, Inc.
3300 Penrose Place, P.O. Box 9140, Boulder, Colorado 80301-9140, USA
www.geosociety.org

Printed in U.S.A.

GSA Books Science Editor: Abhijit Basu

Library of Congress Cataloging-in-Publication Data

Richardson, Keith, 1971
A typology of sculpted forms in open bedrock channels / by Keith Richardson , Paul Anthony Carling.
p. cm. -- (Special paper ; 392)
Includes bibliographic references.
ISBN 0-8137-2392-2 (pbk.)
1. Fluvial geomorphology. 2. River channels. 3. Sediment transport. I. Carling, Paul. II Geological Society of America. III. Title. IV. Special papers (Geological Society of America) ; 392.
.

GB561.R53 2005
551.44′2--dc22

2005049455

Cover: Antelope Canyon, near Page, Arizona. Photographer is Keith Beven, Lancaster Environment Center.

10 9 8 7 6 5 4 3 2 1

Contents

Geological Society of America
Special Paper 392
2005

A typology of sculpted forms in open bedrock channels

Keith Richardson
Lancaster Environment Center, Lancaster University, Lancaster, LA1 4YB, UK

Paul Anthony Carling
School of Geography, University of Southampton, Highfield, Southampton, SO17 1BJ, UK

ABSTRACT

Erosional sculpted forms from open bedrock channels in a variety of lithologies are described, classified, and illustrated. The resulting typology demonstrates both the diversity of bedforms in bedrock channels and the existence of features common to many channels and lithologies. This investigation develops a consistent nomenclature and places the study of bedforms in bedrock channels on a rational foundation along with that of their sedimentary counterparts. Four general principles applying to the development of sculpted forms are identified: (1) the continuity of form between end-member types; (2) the convergence of form toward "preferred" morphologies; (3) the constructive interference of the flow structures of contiguous bedforms; (4) the tendency toward the development of sharp transverse crests in bedrock surfaces. The origins of sharp-crested transverse features, and of sharp edges in sculpted forms in general, are discussed. The use of sculpted forms as indicators of flow patterns, and of the dominant erosion processes active within channels, is examined. It is noted that in channels in silicate rocks with abundant sculpted forms, abrasion by suspended load is the main mechanism of erosion, and that erosion by bedload plays only a minor role. In such channels, lithology per se exerts relatively little influence on the formation of sculpted forms, although structural elements of the substrate do. Substrate structure may influence the style of bedforms and/or inhibit their development. Limestone develops a suite of endemic features, indicative of formation by corrosion. Factors that potentially exert control on the scale of bedforms are discussed.

Keywords: bedrock channels, bedforms, sculpted forms, erosion, channel morphology.

INTRODUCTION

The morphology of surface bedrock channels remains a relatively neglected area of research within the study of fluvial systems (Selby, 1985: p. 262; Ashley et al., 1988; Wende, 1999; Whipple et al., 2000a). There are probably several factors contributing to this state of affairs. Bedrock channel morphology is highly variable, even within a single reach (Karcz, 1973; Tinkler and Wohl, 1998a), and often reflects not just local stream hydraulics, but also rock lithology and heterogeneity, joints, bedding and cleavage, or regional structure and base-level history, depending on the scale of interest (Wohl, 1998). The hydrological regime and climate of the catchment, not to mention biological influences, may also play a role. Given this variability, the morphology of bedrock channels may seem a rather intractable problem. There may be a general view that bedrock channels are not common and are therefore unimportant. In addition, bedrock channels pose considerable difficulties for both field and laboratory studies. Morphological change in bedrock channels is generally extremely slow compared with that in alluvial channels because of the substrate resistance. Furthermore, most erosion is effected in brief, high-discharge events that punctuate extended periods of low flows (Tinkler and Wohl, 1998a), and discharge in bedrock channels tends to be flashy. Therefore, direct observation of morphological change, measurement of erosion rates, or of the

Richardson, K., and Carling, P.A., A typology of sculpted forms in open bedrock channels: Geological Society of America Special Paper 392, 108 p., doi: 10.1130/2005.2392. For permission to copy, contact editing@geosociety.org. ©2005 Geological Society of America.

conditions under which erosion occurs, are difficult. Likewise, simulation of bedrock channel morphology in the flume, unlike that of alluvial channels, cannot use the same material as that found in nature but requires a suitable analogue and is rarely attempted.

Nevertheless, a number of studies of bedrock channels at a variety of scales do exist. At the reach and basin scale, channel planform patterns have been studied, for example, by Bretz (1924) and Baker (1973, 1978), who describe anabranching channel complexes from the scablands of the Columbia Plateau of eastern Washington. Kale et al. (1996), van Niekerk et al. (1999), Heritage et al. (1999), and Tooth and McCarthy (2004) describe anabranching bedrock channels, variously relating channel patterns to the influences of substrate, tectonic activity, and channel gradient. Pohn (1983) examined the relationship between channel orientation and regional joint pattern. Ashley et al. (1988) found that meanders can be found in bedrock channels with a wavelength consistent with that of meanders in alluvial channels. Meandering bedrock channels are also described by Itakura and Ikeda (1997). Montgomery and Gran (2001) studied the variation of bedrock channel width with upstream drainage area, and examined the controls on bedrock channel width and the distribution of bedrock and alluvial reaches within a basin. There have also been many studies on long profiles in relation to tectonic, substrate, and hydraulic controls, and in relation to erosion mechanisms. Kobor and Roering (2004) performed an analysis of upstream drainage area and bedrock channel slope for a large number of basins in the Oregon Coast Range and found a band of elevated channel slopes coincident with the strike of active folds. Righter (1997) invoked base-level lowering to explain the existence of knickpoints in the Atanguillo River, Mexico, while Wohl et al. (1994) found that substrate variability and base-level changes control the distribution of knickpoints in Nahal Paran, Israel. Miller (1991) showed that structural and lithological controls strongly influence the style and distribution of knickpoints in downcutting streams. Wohl and Ikeda (1998) examined the influence of substrate strength and stream power in controlling the magnitude of erosion, knickpoint density, and consequently the form of the long profile. Clear relationships were found by Wohl (2000), Wohl and Merritt (2001), and Wohl and Achyuthan (2002) between the ratio of hydraulic driving forces to substrate resisting forces and the resultant morphology. Seidl et al. (1996) and Weissel and Seidl (1997, 1998) investigated long-profile development and knickpoint retreat at a passive continental margin and inferred that mass wasting at the knickpoints coupled with fluvial processes are necessary to explain the observed rate of retreat. Erosion rates in natural bedrock channels have been measured by Foley (1980a), Smith et al. (1995), Righter (1997), Hancock et al. (1998), Tinkler and Parish (1998), Whipple et al. (2000b), Wohl and Achyuthan (2002), Kale and Joshi (2004), and Reusser et al. (2004). Foley (1980b), Sklar and Dietrich (1998), Izumi and Parker (2000), and Parker and Izumi (2000) have constructed mechanistic mathematical models of erosion in bedrock channels.

At the subreach scale, a range of morphological features has been described. Large-scale scoured features such as pool-riffle and step-pool sequences similar to those found in alluvial channels have been reported by several studies (Keller and Melhorn, 1978; Baker and Pickup, 1987; Wohl, 1992a, 2000; Wohl and Grodek, 1994; Duckson and Duckson, 1995; Wohl and Ikeda, 1998; Wohl and Merritt, 2001). Other large-scale features include inner channels and bedrock benches (Baker, 1973, 1978; Baker and Pickup, 1987; Wohl, 1992a, 1992b, 1993; Wohl et al., 1994; Gupta et al., 1999; Wohl and Achyuthan, 2002) and undulating canyon walls (Wohl et al., 1999; Wohl and Achyuthan, 2002). A range of large-scale erosion features is associated with the Channeled Scabland of eastern Washington, produced by catastrophic paleofloods (Baker, 1973, 1978). Smaller-scale sculpted forms include the often ignored but important smoothed and polished rock surfaces (Baker and Pickup, 1987; Tinkler, 1993; Baker and Kale, 1998; Hancock et al., 1998; Wohl and Ikeda, 1998; Gupta et al., 1999; Spaggiari et al., 1999; Whipple et al., 2000a), ripple-like features (Sevon, 1988; Wohl, 1992b; Hancock et al., 1998; Whipple et al., 2000a), potholes (King, 1927; Alexander, 1932; Ives, 1948; Ängeby, 1951; Blank, 1958; Gjessing, 1967; Morgan, 1970; Allen, 1982: p. 261–264; Nemec et al., 1982; Baker and Pickup, 1987; Sato et al., 1987; Wohl, 1993; Lorenc et al., 1994; Zen and Prestegaard, 1994; Tinkler, 1997a; De Wit, 1999; Jacob et al., 1999; Spaggiari et al., 1999; Whipple et al., 2000a; Kunert and Coniglio, 2002; Wohl and Achyuthan, 2002), and a wide range of other scoured depressions such as flutes and grooves (King, 1927; Maxson and Campbell, 1935; Blank, 1958; Allen, 1971a; Selby, 1985: Plate 9.4 therein; Baker and Pickup, 1987; Wohl, 1992b, 1993; Hancock et al., 1998; Gupta et al., 1999; Whipple et al., 2000a; Wohl and Achyuthan, 2002). Several studies have described the importance of bedrock channels in general, and bedrock sculpted forms in particular, in providing fixed trap sites of high turbulence and slow aggradation for the accumulation of mineral placers (e.g., De Wit, 1999; Jacob et al., 1999; Spaggiari et al., 1999).

In other environments and materials too, sculpted features have been extensively studied. This is particularly true of structures known as scallops found in caves (Coleman, 1949; Curl, 1966; Goodchild and Ford, 1971; Wigley, 1972; Blumberg and Curl, 1974; Bögli, 1981; Lauritzen et al., 1983; Gale, 1984; Slabe, 1996; Murphy et al., 2000; Springer and Wohl, 2002), and karst features known as karren (Sweeting, 1972: p. 74–102; Allen, 1982: p. 223–248; Bögli, 1981; Trudgill, 1985: p. 53–62; Ford and Williams, 1991: p. 375–396). In glaciated areas, a suite of bedforms known as s-forms has been identified, and a well-established terminology developed (Kor et al., 1991; Tinkler and Stenson, 1992; Shaw, 1996; Kor and Cowell, 1998). These features are thought to be formed by the catastrophic release of subglacial meltwater, and similar erosion patterns have also been described from a bedrock channel (Tinkler, 1993). Morgan (1970) described potholes in bedrock buried by glacial deposits and interpreted them as a subglacial meltwater channel. Bryant (2001: p. 82–89) described sculpted forms in coastal areas due to tsunamis. Rippled surfaces on the underside of river ice covers have also been studied (Ashton and Kennedy, 1972; Epstein and Cheung, 1983). Considerable literature is devoted to the study of aeolian sculpted forms in bedrock (e.g., Maxson, 1940; Sharp, 1949; Selby, 1977; Summerfield, 1991; Schlyter, 1994; Livingstone and Warren, 1996; Braley and Wilson, 1997), and to a much smaller extent in snow, ice, and sand (e.g., Allen, 1965; Curl, 1966; Whitney, 1983). Aeolian features often show similarities to fluvial scoured forms but are outside the scope of this report.

As mentioned above, there have been few experimental studies of bedrock channel morphology. A seminal work is that of Allen (1971a), who generated flutes and scallops from defects by dissolution in plaster of Paris. Goodchild and Ford (1971) and Blumberg and Curl (1974) studied scallop wavelength and scallop profile evolution respectively in the same medium. Other workers have used cohesive substrates consisting of sand-clay mixtures. Shepherd and Schumm (1974) and Kodama and Ikeda (1984) studied the development of inner channels in cohesive substrates, while Gardner (1983) studied knickpoint and longitudinal profile evolution. Wohl and Ikeda (1997) and Koyama and Ikeda (1998) have examined the effect of gradient on channel morphology.

However, despite these numerous studies of bedrock channels, naturally occurring examples of bedforms remain poorly described. This is especially true at the smaller end of the scale. Several workers have described the small-scale erosional features of their field sites, but only as indicators or illustrations of the general principles of active processes (e.g., Wohl, 1992b, 1993; Baker and Kale, 1998; Hancock et al., 1998; Wohl and Ikeda, 1998; Whipple et al., 2000a). Studies that do focus on the sculpted forms themselves consider only a limited range of bedform types (e.g., Elston, 1917, 1918; Alexander, 1932; Maxson and Campbell, 1935; Nemec et al., 1982; Sato et al., 1987; Zen and Prestegaard, 1994; Duckson and Duckson, 1995), or a limited range of sites, such as a short stretch of a single river (King, 1927; Ängeby, 1951; Blank, 1958; Sevon, 1988; Tinkler, 1993). A holistic approach has never been attempted. Furthermore, published descriptions are brief, lack sufficient illustrations, and show the existence of a confusing, nonstandardized, and ill-defined terminology. This means that features described in the literature cannot properly be compared with sculpted forms found by investigators in the field, a prerequisite to the scientific study of them. Most importantly, however, fieldwork undertaken during this study has shown that there is a wealth of small-scale bedforms in bedrock channels, which does not feature in the existing literature.

This review describes a wide range of sculpted features in open bedrock channels (i.e., excluding cave passages) from a variety of lithologies as a first attempt at placing the study of bedrock bedforms on a rational foundation comparable with the study of their alluvial counterparts. The very large-scale morphological features of bedrock channels, such as channel planform, longitudinal profile shape, and terraces, are beyond the scope of this study. Instead, it will focus on small-scale (millimeters to meters) forms, but with mention of features at the scale of the channel cross section. A large section of this report (i.e., the actual typology) is given over to detailed and methodical descriptions of bedforms, and this is intended to be reference material to assist in the identification of bedforms. Other sections of this report (those preceding and following the typology section) are of more general interest and discuss mechanisms and underlying principles of the development of sculpted forms.

A great unknown in the study of bedrock channels is that of the detailed flow structures associated with the majority of sculpted forms. Although this report focuses on bedform morphology and does not consider associated flow structures in detail, a basic assumption that underlies the typology presented here is that, in a given hydrodynamic environment, the form of the bedrock surface controls the flow patterns present above and downstream of it. Because the flow structures are the agents of erosion, this implies that the detailed morphology of bedforms both determines and is influenced by the flow structures in a feedback loop. It also implies that the morphology, if we are able to interpret it correctly, is an indicator of the flow patterns, and that features of different morphologies have different associated flow structures. It follows that the diversity of bedforms and the description of that diversity are of great significance in understanding flow in bedrock channels, and that such a description is a prerequisite to a full exploration of flow structures, such as that carried out experimentally for flutes by Allen (1971a). This consideration is a further motivation for producing this typology.

General Remarks on Bedforms in Open Bedrock Channels

In spite of the analogies that can be drawn between bedrock and alluvial bedforms, there are several fundamental differences between them, and it is worth considering these similarities and differences before proceeding to the details of individual features. Throughout this report, the names of sculpted forms included in the typology, and of channel morphologies, are italicized.

(i) In common with some sedimentary scour marks, the evolution of bedrock bedforms is a unidirectional process involving only erosion (Allen, 1971a; Tinkler and Wohl, 1998a), rather than the interaction of erosion and deposition as in most sedimentary bedforms. Because bedrock sculpting is also a slow process (compared with the lifetime of a sedimentary bedform), local velocity fields within the channel can be considered permanently modified (Tinkler and Wohl, 1998a) due to the feedback between flow structures and bedforms, at least over short to moderate time scales. Over long time scales, such as the recurrence interval of major floods, some authors have suggested that even bedrock channels should be considered to have deformable boundaries, which adjust to the flow in a manner analogous to alluvial channels (e.g., Keller and Melhorn, 1978; Ashley et al., 1988; Wohl, 1992a; Wohl and Grodek, 1994; Grant, 1997; Baker and Kale, 1998).

(ii) Both bedrock and alluvial bedforms may exist as isolated individuals, as patterns of contiguous individuals with shared boundaries, or as a train of bedforms (i.e., in the manner of ripples and dunes).

(iii) Bedrock bedforms may develop in one of five ways:
- *Concave features* may erode at approximately equal rates in all directions (seen in plan view), the centroid remaining in the same position, for example, some *potholes*. Springer and Wohl (2002) describe subspherical *incipient lateral potholes*, which they infer remain locked in position, and which maintain constant form as they grow.
- *Concave features* may erode more rapidly in one direction than in other directions, resulting in the migration of the centroid in that direction, for example, *flutes* (Allen, 1971a).
- After an initial phase of erosion and mutual adjustment, a train of bedforms may either (1) migrate as a stable profile without individuals changing in size or shape (e.g., certain stable types of *scallops* [Curl, 1966; Blumberg and Curl, 1974]), (2) exhibit stability only in a statistical sense as a collective, and migrate with simutaneous nucleation, growth, and destruction of individuals (e.g., *scallops* in general [Allen, 1971a]), or (3) exhibit neither of these forms of stability, with both individuals and the statistical properties of the population changing, perhaps as a response to changed hydraulic conditions (Blumberg and Curl, 1974).

(iv) In terms of erosion, bedrock acts as a medium- and high-pass filter (Tinkler and Wohl, 1998a). In other words, the erosion recorded in the bedrock is a temporal averaging of the effect of many individual erosive events. These events may be presumed to be intermittent in nature as a result of the intermittent character of the turbulent coherent structures to which they are related and the intermittent nature of high-discharge events, which are the most geomorphologically effective (Tinkler and Wohl, 1998a). Alluvial bedforms also exhibit a degree of temporal averaging; however, the importance of this averaging increases with the size of the feature in proportion to the amount of sediment it contains and the time required to transport that sediment. In contrast, the temporal averaging exhibited by sculpted forms in bedrock is a result of the inherent strength of the substrate and is far greater than that shown in sedimentary features of a similar size.

(v) The development of bedrock bedforms is the result of marked spatial variation in erosion rates on a scale much less than the channel width. This is equally true of alluvial bedforms, with the proviso that usually spatial variation in both erosion and deposition rates is involved. In the case of a train of bedforms (such as *scallops* and *pseudo-ripples* in bedrock), the spatial variation in erosion rates is quasi-periodic in nature.

(vi) As a consequence of (i) and the fact that there is normally no mobile bed of sediment, gravitational forces are irrelevant to all bedrock bedforms except those that require interaction with a free surface, those in which abrasion by bedload plays a part, and those covered by low-stage flow when sculpted forms may act as topography (see point ix). Thus, most bedforms in bedrock can occupy surfaces of any orientation. This is in contrast to most alluvial bedforms, which occur only on horizontal and near-horizontal surfaces because they involve both erosion and deposition and interaction with a bedload.

(vii) As a result of the relatively slow response of bedrock channels, sculpted forms, which are thought to be actively formed mainly in very high flows under high stream power by water with large amounts of suspended sediment, persist through a range of lower flows (Baker and Kale, 1998; Tinkler and Wohl, 1998a) and periods of subaerial exposure. Many erosional forms in bedrock are in fact relict features, left above the level of floodwaters due to channel incision (King, 1927; Jennings, 1983; Wohl, 1999). This contrasts with sedimentary bedforms such as antidunes and upper-stage plane beds, which are rarely preserved because they evolve into other bedforms as the flow reduces in strength. Dunes will persist through lower flows than those in which they are active because of their large size, although even in this case the bedforms are often degraded by the lower flow. In addition, small-scale sedimentary bedforms such as ripples are quickly destroyed by subaerial exposure. It is, however, only during low flows and subaerial exposure that bedrock bedforms can be observed; that is to say they are observed not under their formative conditions, but only when they are inactive. Thus, the flow patterns within the channel at the time of observation are normally no guide to the mode of formation and development of sculpted forms. Indications that sculpted forms in bedrock are active, rather than relict, forms include a lack of signs of chemical weathering, a lack of biological growth (lichen, moss, etc.) and the presence of sharp edges and freshly abraded or polished surfaces (Wohl, 1999).

(viii) By comparison with alluvial channels, the boundaries of most bedrock channels are irregular and hydraulically very rough. The irregularities, which range in size from the millimeter scale to that of the depth of the flow, generate recirculating eddies of various sizes and orientations on both their downstream and upstream sides and cause flow to diverge and converge around bedrock highs. This leads to spatially complex flow patterns and highly non-uniform flow, and mean flow vectors locally can deviate significantly from the trend of the channel. Sculpted forms in bedrock channels therefore often form in response to very local flow patterns and hydrodynamic conditions, and their orientation may not reflect the general direction of the channel. *Entry furrows* do not always occur on the upstream side (with respect to the main channel) of *potholes*, for example. This is a disadvantage when one is trying to understand the flow dynamics of individual bedforms, because one cannot say a priori which will be the upstream side of the feature when it is active.

(ix) Flow patterns in river channels generally change with stage (e.g., Whiting, 1997), and in bedrock channels the change is possibly greater than in alluvial channels as a result of their topographical complexity. Thus, flow patterns in bedrock channels are temporally as well as spatially complex. This means that an individual bedform will often experience incident flow from different directions during the hydrograph of a flood event. The development of such a bedform will depend on which flow direction is dominant in terms of erosion, taking into account the erosion intensity of the flow and its frequency of occurrence. However, once all local topography is submerged to a sufficient depth at some moderate to high stage, flow patterns are unlikely to change qualitatively with further increases in stage, although there will be increases in velocity and turbulence intensity. Based on observations made for this study, it is hypothesized here that in general, sculpted forms develop in response to incident flow from only one direction, this being the flow they experience at and above some critical discharge above which the local flow pattern in the channel does not change qualitatively. There are exceptions to this rule. Some bedforms, such as *irregular sculpting*, lack a clear orientation and apparently develop in highly turbulent regimes in response to multidirectional flow. Also, at very low stage, some bedforms in the channel floor (e.g., *runnels*) may begin to act as topography rather than as roughness, and so act to divert water into or around themselves, creating spatially restricted but high velocity flows. Small-scale sculpted forms can develop at very low stage in such flows.

(x) Sculpted forms in bedrock are highly dependent on local, contingent circumstances, such as the local flow patterns and hydrodynamic conditions, the orientation of the bedrock surface with respect to the flow, the local rock structure, and the feedback between these and the developing bedform. That is to say, each sculpted form is a singular feature; it possesses enough characteristics in common with some other bedforms for it to be compared and classified with them, but it also possesses certain traits that set it apart from all others and make it behave or appear differently (see Schumm, 1991: p. 75–76, for a discussion of singularity in geomorphology). This creates an almost infinite variety of forms in bedrock erosional features; the variety is, in fact, as wide as the number of bedforms in existence.

(xi) The variety of morphology in bedrock channels noted above is made possible by the strength of the substrate, which both allows for complexity of form (e.g., steep slopes, overhangs, ridges, projections) and preserves a long history of active erosion. This situation contrasts with that of sedimentary bedforms, which are restricted to development on near-horizontal surfaces, and which are composed of sediment with low or zero mechanical strength (resulting in the maximum slope obtainable being the angle of repose) and relatively consistent and uniform properties. Under these circumstances, bedform morphology is more uniform, and bedform types may be predicted from relatively few parameters, such as grain size, and recent flow depth and shear stress history.

TERMINOLOGY AND DEFINITIONS

Introduction

The nomenclature previously used to describe forms in open bedrock channels is varied and nonstandardized. Therefore, a clarification and revision of some commonly used expressions and existing nomenclature is necessary. The definitions and nomenclature detailed below are those adopted in this report and are suggested as potential standards. These standards are presented as an aid in separating an often bewildering array of bedforms. We acknowledge that once further hydraulic flume tests and computer modeling are completed for some bedform types, it is entirely possible that the typology will be in need of revision.

Erosion Mechanisms

There are several accounts of the mechanisms of fluvial erosion in bedrock channels, and it is not necessary to describe the processes in detail here. The interested reader is referred to the useful works of Allen (1971a), Selby (1985: p. 242–245), Hancock et al. (1998), Wohl (1998), Wende (1999), and Whipple et al. (2000a). However, brief definitions of the various processes are provided here for reference. These definitions are not revisions of existing terminology.

(i) Plucking, also known as quarrying and jacking, involves the removal of whole blocks of rock from the boundary by lift and drag forces acting on the block. The blocks are delineated by joints and other structural features of the rock and undergo a period of preparation prior to their entrainment by the flow. Block preparation involves the widening and propagation of cracks and the loosening of blocks by a combination of hydraulic forces, bedload impact, wedging of sediment within cracks, and physical and chemical weathering (Whipple et al., 2000a).

(ii) Abrasion, also known as corrasion, involves the wearing away of a surface by numerous impacts from sediment particles transported in the fluid. These particles may be carried in suspension or as bedload, or both. Each impact, if sufficiently energetic, breaks off a small piece of the boundary material, which may be at the scale of a small part of a grain in the case of suspended load, or as much as a substantial flake or chip of rock in the case of coarse bedload.

(iii) Fluid stressing, also known as evorsion, refers to the removal of particles by the turbulent stresses exerted directly on the boundary by a clear fluid. At the scale of individual grains and grain aggregates, this process is likely to be important only in clays and poorly consolidated rocks. However, at larger scales, this mechanism is important in flaking, and it is the process by which prepared blocks are plucked.

(iv) Cavitation occurs in very rapid flows when the instantaneous dynamic fluid pressure locally drops below the vapor pressure and bubbles of water vapor appear. The water effectively boils. The bubbles very rapidly collapse, however, and the pressure shock wave thus generated erodes any nearby boundary. Cavitation is known to occur on engineering structures (Arndt, 1981), but whether it occurs in natural channels is still a matter of debate (Baker, 1973; Sato et al., 1987; Wohl, 1992b; Baker and Kale, 1998; Hancock et al., 1998; Tinkler and Wohl, 1998a; Gupta et al., 1999; Whipple et al., 2000a).

(v) Corrosion occurs on soluble rocks, where it is also known as dissolution, and on chemically reactive rocks, where it is referred to as chemical weathering. Soluble rocks include limestone, marble, and evaporites. Other rocks, such as quartzites, extrusive igneous rocks, granitic rocks, and rocks with a high calcareous content such as marl, are slightly soluble or have soluble components and may also show evidence of dissolution. Chemically reactive rocks have unstable components, such as feldspars, olivine, and sulfide minerals, which break down into various decay products and are subsequently easily removed. Corrosion may occur subaqueously or subaerially. It acts both directly in removing the rock or components of it, and indirectly in physically weakening the rock and assisting its removal by the other erosion mechanisms. It is possible that the growth of algae, moss, lichen, etc., on the rock enhances corrosion by inhibiting evaporation and maintaining a film of water in contact with the rock when surfaces elsewhere are dry.

(vi) Physical weathering acts indirectly in fluvial erosion by physically weakening the rock (Whipple et al., 2000a) and increasing its surface area as a result of crack propagation, thereby enhancing erosion by the mechanisms listed above. Perhaps the most important physical weathering processes in bedrock channels are wetting and drying cycles, weathering due to the crystallization of salts (Sparks, 1972: p. 28–29), and freeze-thaw cycles, all of which play a role in crack propagation, block loosening, and rock disaggregation. Blocks or layers of rock can also be removed directly by ice lift, when a layer of ice forms on the rock surface during low flow in winter and rises to the surface during a subsequent flood, carrying with it small attached blocks of rock (Tinkler and Parish, 1998; see also Carling and Tinkler, 1998, for a discussion on the effect of ice in increasing the transportability of loose blocks). Laminated rocks and foliated rocks such as shales are often easily disaggregated by the process of flaking and are particularly susceptible to the physical weathering mechanisms described above (Tinkler and Parish, 1998).

A Definition of Bedforms in Open Bedrock Channels

The most fundamental definition is that of the range of features that comprise fluvial bedrock bedforms. This definition should be restrictive in the sense that not every topographic detail of a channel can be considered as a bedform. Thus, it must allow bedforms so defined to be distinguished from

(i) features that result from processes other than fluvial ones

(ii) surfaces that might exist even if the rock outcrop were outside a fluvial environment

(iii) the general "morphological noise" that exists on any rock outcrop

The first two groups of surfaces listed above are closely related and can be discussed together. In very steep channels (headwaters, waterfalls, etc.) and channels in gorges, and on bedrock surfaces that are inundated only intermittently, fluvial processes compete with subaerial physical and chemical weathering (Howard, 1998; Tinkler and Parish, 1998; Whipple et al., 2000a), karstification (King, 1927), and hillslope processes such as debris flow and slope failure (Howard, 1998; Sklar and Dietrich, 1998; Weissel and Seidl, 1998). The morphological products (in bedrock surfaces) of such processes may or may not be readily identified, but as a general point it is suggested here that features arising from nonfluvial processes should not be included as bedrock channel bedforms. Features of the second group may result from the nonfluvial processes considered above but can also result from fluvial processes (i.e., plucking and flaking), which is why they are separated from the first group. Included are surfaces arising from structural effects (joint, bedding, and cleavage surfaces, etc.) and from lithological components (intraclasts, fossils, sedimentary structures, large grains, etc.). They also include forms resulting from diagenetic effects (such as concretions, chemical alteration by groundwaters, and recrystallization). The existence of such features depends primarily on the intrinsic rock properties. These features may therefore be present on rock surfaces in a range of environments and are not peculiar to fluvial environments.

The last group of features includes the effectively random forms present on a rock surface arising from the chance interaction of several influences, such as the detailed internal structure of the rock, its local diagenetic history, its erosion history within the channel, subaerial weathering effects, and any biological influences. Such features are beyond "explanation" because their presence is due to one or more chance events. They are not identifiable individually as distinctive features, unlike those described above, but collectively they constitute an *irregular surface*. The idea of morphological noise is a more important concept in the study of bedrock channels than alluvial channels, because a mass of sediment retains less of its history and responds more immediately and directly to hydrodynamic influences.

In the light of this discussion, in this report, fluvial bedforms are considered to be those forms resulting principally from sculpting of the rock by any fluvial mechanism (abrasion, corrosion, fluid stressing, and cavitation), producing smoothed and rounded surfaces that do not primarily reflect the intrinsic properties of the rock. Such surfaces must be clearly distinguishable from a simple rough or irregular rock surface. The plucking of bedrock blocks, and flaking in shaly rocks, although fluvial processes, expose structural surfaces (Whipple et al., 2000a) that are intrinsic to the substrate rather than to the agent of erosion and could be present in a range of environments. Furthermore, such surfaces cannot be classified except as a type of geological structure. Plucking is therefore not considered directly to produce fluvial bedforms (an exception to this is very large scale features; see definition of *potholes*, below), although surfaces thus exposed are subsequently available for scouring and bedform development. Bedrock bedforms thus defined are analogous to sedimentary bedforms in terms of their hydrodynamic control, and they are genetically identical to the class of sedimentary bedforms known as scour marks or erosional marks (e.g., Dzulynski and Sanders, 1962; Allen, 1971a; Karcz, 1973; Collinson and Thompson, 1982: p. 36–45). In this report, the terms scour mark, sculpted form, and (bedrock) bedform are treated as synonyms.

Describing the Effect of Rock Structure on Channel Morphology

Structural versus Hydrodynamic Control of Morphology

The above definition of fluvial bedforms poses the question of how one should consider a bedrock channel in which plucking dominates the erosion process, and whose boundary is composed of exposed joints or bedding planes, because according to the above definition such a channel does not contain bedforms. Such channels have been described by Bretz (1924), Pohn (1983), Miller (1991), Wende (1999), Whipple et al. (2000a) and Tooth and McCarthy (2004: Fig. 8B therein). A broad classification of channel morphological types into "structurally controlled" type (heterogeneous or structured substrate dominated by plucking and flaking) and "hydrodynamically controlled" type (massive substrate dominated by abrasion, corrosion, etc.) might be appropriate. Thus, structurally controlled bedrock channels contain few, if any, bedforms; the boundary consists almost entirely of surfaces of structural origin, and the channel morphology follows structural trends (Fig. 1A). The efficacy

Figure 1. A: Structural control of channel morphology. This channel is dominated by structural surfaces. A vertical cleavage intersects with bedding planes dipping steeply to the right of the photograph, forming upward-pointing projections ending in acute angles. River Stinchar, UK, looking upstream. Mudstone. Person for scale. B: Structural control of reach orientation. The steeply sloping surface on the right is a bedding plane. River Eden, UK, looking upstream. Interbedded sandstone, limestone, and shale. The channel is ~8 m wide.

of plucking and flaking depends on the degree of anisotropy in the rock fabric and the spacing of fractures in the rock, but Whipple et al. (2000a) suggest that where joints are spaced less than 1 m apart, plucking will be dominant, and abrasion will be insignificant. Plucking also depends on the number of joint and cleavage sets and their orientation. Pohn (1983) argued that the orientation of streams over bedrock is independent of that of joints in the substrate, but in fact there is often a strong control exerted by joints, and by steeply dipping bedding planes, on stream orientation and planform, at least over short reaches (Fig. 1B; see also Tooth and McCarthy, 2004).

The boundaries of hydrodynamically controlled channels, on the other hand, consist mainly of bedforms and have few surfaces of structural origin or following a structural trend. However, the question of scale is important when considering bedrock structure, or more specifically, the scale of defect and structure size and spacing compared with the dimensions of potential bedforms. For example, while a conglomerate may be regarded as massive at the reach and cross section scale, at the scale of the sculpted form, it represents a complex system of internal surfaces and large defects that interrupt and prohibit the growth of bedforms. A channel in such a rock may therefore be hydrodynamically controlled at the reach and cross section scale, but structurally controlled at the scale of the sculpted form, and therefore lacking in small-scale bedforms.

A common form of structurally controlled channel is the dip-parallel reach (Miller, 1991), in which the dominant structural surfaces (typically bedding planes) have approximately the same slope as the channel, and water flows over individual rock strata for some distance. For slight deviations of the dip of the bedding planes from the general channel slope, channel reaches are punctuated by *steps*, which are either *forward-facing* (downdip reach) or *backward-facing* (updip reach), depending on the relative slope of the bedding planes and the channel (Miller, 1991; Wende, 1999). Such channel morphologies are here termed *concordant plane beds* (Figs. 2A and 2B) and represent channels exhibiting structural morphological control. The channel floor is often very smooth and almost devoid of sculpted forms. *Concordant plane beds* are described by King (1927), Blank (1958), Miller (1991), and Wohl (1993). In

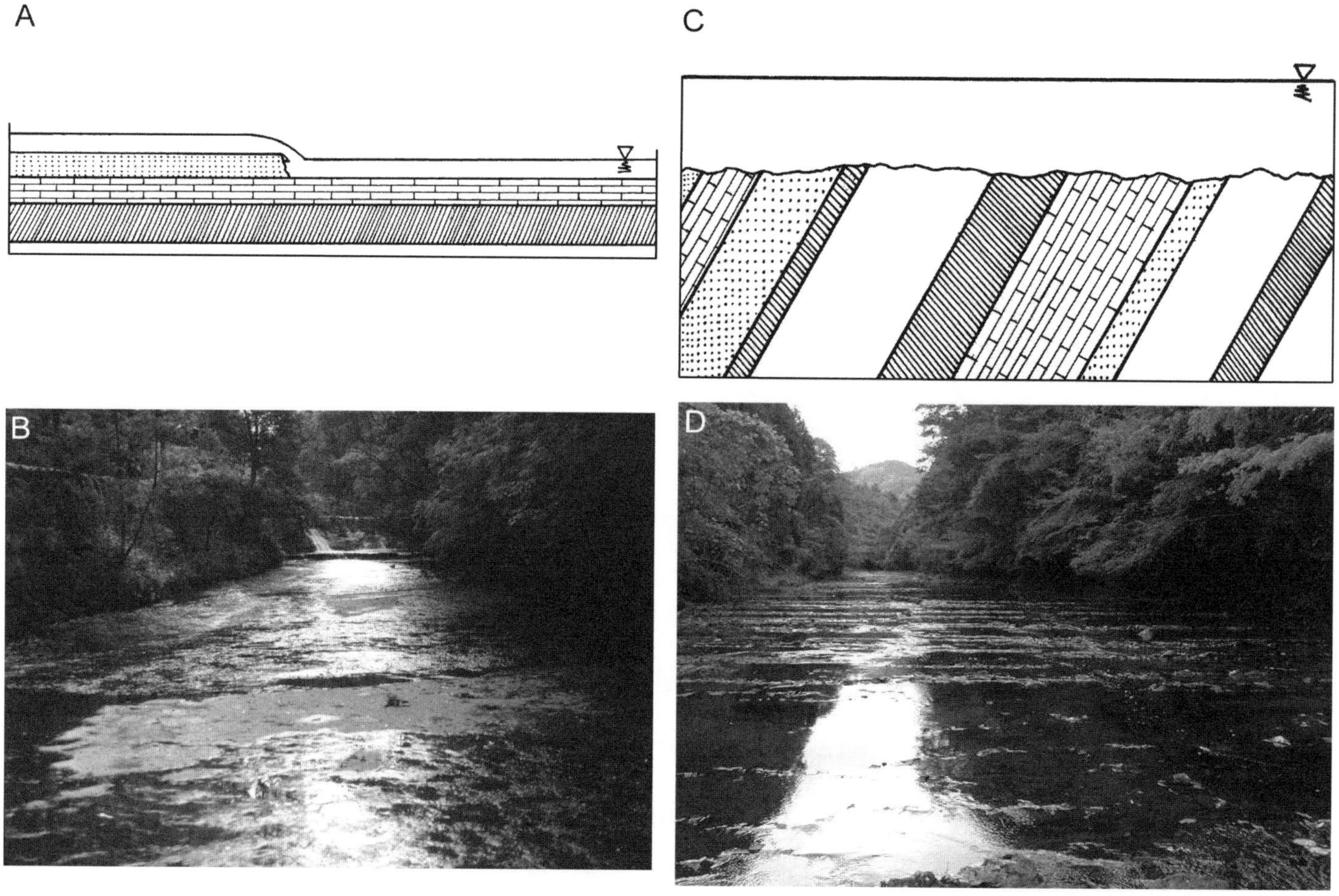

Figure 2. A: Longitudinal cross section through a *concordant plane bed*. B: Channel with a *concordant plane bed*. The channel floor consists of a single bedding plane. River Dee, UK, looking upstream. Limestone. The channel is approximately 10 m wide. C: Longitudinal cross section through a *discordant plane bed*. D: Channel with a *discordant plane bed*. The ridges running across the channel are formed by alternating hard (sandstone) and soft (mudstone) beds striking across the river and dipping steeply. Sichiri-gawa, Japan, looking upstream. The channel is ~30 m wide.

contrast, some channels have planar beds in which the main structural surfaces of the rock are at a high angle to the channel floor. Such channel boundaries may be termed *discordant plane beds* (Figs. 2C and 2D; see also section 6.1a). In this case, although the rock structure may exert morphological control at the level of small-scale sculpted forms, at the macroscopic (i.e., channel width and reach) scale, it does not. *Discordant plane beds* are planar in spite of, rather than because of, the structure of the substrate, in contrast with their concordant counterparts, and represent channels in which large-scale structural control of morphology is absent. It is worth noting that the term *plane bed* is used in the macroscopic sense; in detail, both varieties of *plane beds* may have surfaces showing a relief of several decimeters. For example, the channel shown in Figure 1A could be said to have a *plane bed* at the macroscopic scale.

Another type of structurally controlled channel morphology is an *irregular surface*. Such surfaces may be formed within channels in highly heterogeneous rock, rock with a large concentration of "defects" (e.g., fossils, concretions, intraclasts, veins, and cavities), in especially coarse-grained lithologies (conglomerates, breccias, agglomerates, etc.), and in rocks with irregular structural surfaces. *Irregular surfaces* are also common toward channel margins, where fluvial sculpting processes are weak or intermittent, and compete with subaerial weathering or hillslope processes.

Structurally Influenced Channel Morphologies

It is rare to find rock outcrops that have no structure whatsoever, and depending on the degree of anisotropy in the rock fabric and on the spacing of fractures, structural elements impose a greater or lesser influence on channel morphology. Thus, between the two end-member types of hydrodynamically controlled and structurally controlled channels is a range of channel morphologies with varying degrees of structural control and modification of surfaces by abrasion, which might be appropriately termed "structurally influenced." In structurally influenced channels, the effects of both plucking and abrasion (or corrosion, etc.) are significant. The structural influence is manifested at both reach and cross section scales, and in the location, orientation, and morphology of individual sculpted forms. For instance, Wohl (2000), Wohl and Merritt (2001), and Wohl and Achyuthan (2002) proposed that bedrock channel morphology reflects the controlling variables of hydraulic driving forces and substrate resisting forces. Wohl (2000) showed that the slope of bedding planes relative to channel gradient (i.e., whether the reach is updip or downdip) has a strong influence on channel morphology, for example, the *step* height:length ratio and *pool* volume in *step-pool* sequences, as a result of the effect the relative slope has on the resistance of the substrate. Wohl and Merritt (2001) developed a model that predicted reach-scale and cross-section-scale channel morphology types based on a combination of variables describing hydraulic driving forces and substrate resisting forces, and that correctly classified 70% of the reaches used in the study. These authors did not consider small-scale sculpted forms, however, or the detailed manner in which structural influences are exerted on channel morphology.

The interaction of sculpting processes with structural, lithological, and diagenetic features was discussed by Elston (1918) in his study of *potholes* and other features. It produces "valid" bedforms provided there are clear signs of modification by sculpting, i.e., that a significant proportion of the rock surface is abraded and is composed of surfaces other than those directly of structural, lithological, weathering, and diagenetic origin. In structurally influenced channels, plucked surfaces and internal rock structures present an array of "defects" suitable for the initiation of sculpted forms by abrasion, and lines of weakness for the sculpted forms to exploit. In this way, the intrinsic properties of the rock can exert a strong influence on the style and spatial distribution of bedforms.

Structurally influenced sculpted forms may be classified according to the type of structures they follow, which broadly speaking are of two types:

(i) *Sculpted forms influenced by internal structures.* Internal structures include primary features, such as bedding and fossils (in sedimentary rocks) and flow banding (in lavas), and secondary (diagenetic and metamorphic) features such as concretions, cleavage, and gneissicity. Such internal structures represent discontinuities in the physical properties of the substrate and may be picked out by differential rates of erosion, creating "defects."

(ii) *Concave features influenced by fractures.* Fractures in the rock include joints, brittle shear zones, faults, veins, and dikes.

Both internal rock structures and fractures may influence bedform development in either or both of two ways: firstly by acting as defects that generate attached vortices and therefore initiate bedforms and/or enhance their rate of growth in a particular direction, and secondly by acting as planes of weakness for bedforms to exploit. Sculpted forms, whether longitudinal or transverse in character, would, under purely hydrodynamic control, be orientated with respect to the dominant flow direction but, as a result of the second type of influence, are often modified to be orientated parallel to the structural trend of the rock. The result of the first type of influence is that those bedforms which rely on some form of defect for their initiation, such as *flutes*, *transverse furrows*, and *obstacle marks*, are mostly, but not always, structurally influenced. However, the influence of rock structure is not restricted to these features and can be present in any bedform whatsoever. The most common types of structurally influenced bedforms are *furrows* that follow the lines of joints or bedding planes (Fig. 3). These are referred to as *joint furrows* and *bedding plane furrows* respectively. However, other types of *concave features* are also found to be structurally influenced. For example, *potholes* may be found that have *exit furrows* parallel to the substrate structure (Fig. 4), or they may be found to be elongated not in the direction of flow, but in the direction of cleavage or some other structural trend. *Short furrows* are also

Figure 3. *Bedding plane furrows*. Borrow Beck, UK. Flow from bottom left to top right. Calcareous mudstone. The scale is 60 cm long.

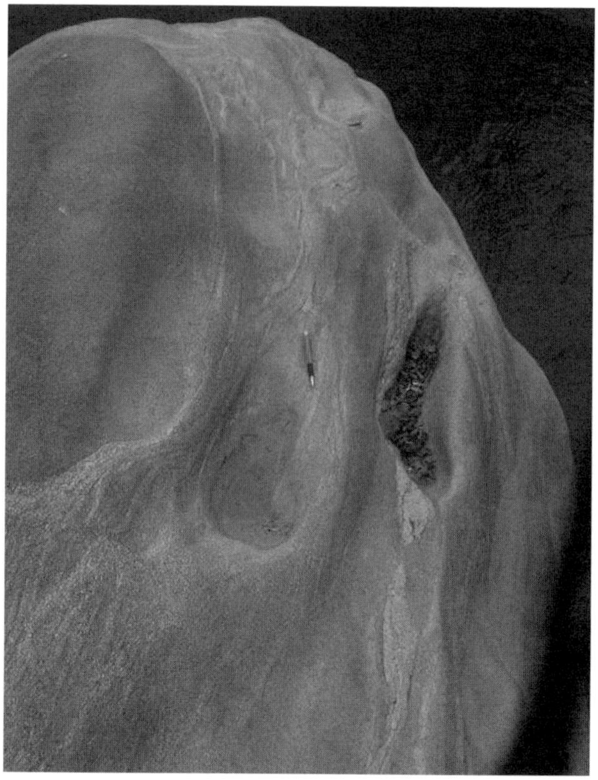

Figure 5. *Structurally influenced short furrows*. These bedforms are on the lee face of a rock projection and are orientated parallel to the gneissic banding in the rock. Nam Mae Chaem (Ob Luang), Thailand. Flow from left to right. Granitic gneiss. Pen for scale.

Figure 4. *Structurally influenced pothole*. This example has an *exit furrow* (bottom left of *pothole*; see section 1.1b of the typology) parallel to the rock's cleavage. River Lune (Tebay), UK. Flow from top to bottom. Calcareous mudstone. Penknife for scale.

sometimes found orientated along structural trends (Fig. 5). *Solution pits* may be found aligned along bedding planes, and individuals may be elongated parallel to bedding. The crests of *sharp-crested hummocky forms* may be parallel to a structural trend (Fig. 157), rather than be transverse to the flow direction. In all such cases, the prefix *structurally influenced* may be applied to bedform names. *Partially abraded surfaces*, in which originally angular, structural elements are partially smoothed and rounded, are also examples of structurally influenced bedforms.

Further Definitions

Simple, Compound, and Coalesced Forms

These terms are sometimes used in the names of bedforms in this study and therefore appear in italics where they are so used. Bedforms of all three of the types below are regarded as individuals, whatever their origin or internal morphology.

(i) "Simple" is used to describe features that are single entities of uncomplicated form and that have developed in isolation from other bedforms (Fig. 6A).

(ii) "Compound" is used to describe bedforms with a more complex internal morphology consisting of a system of

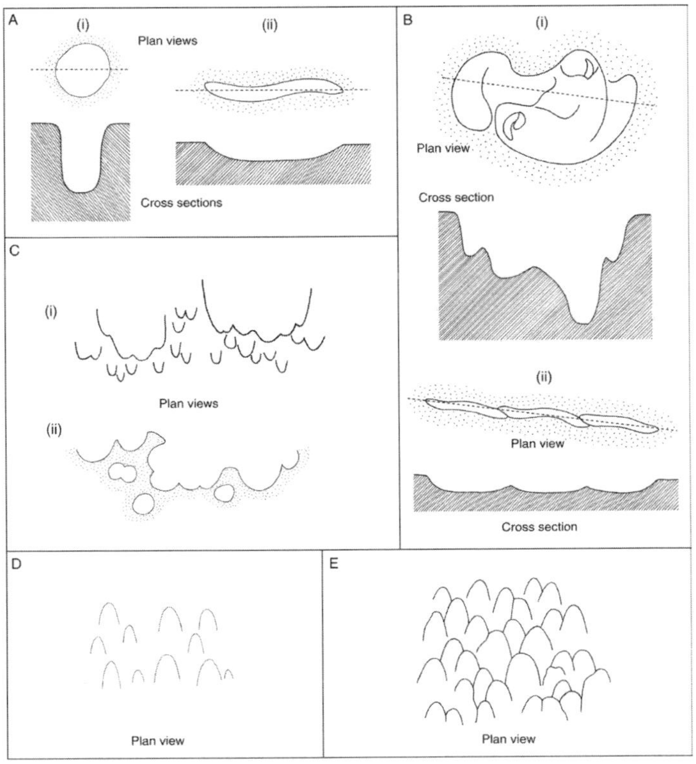

Figure 6. Cartoons illustrating bedform terminology described in the text. A: Simple features: (i) *pothole*, (ii) *longitudinal furrow*. B: Compound features: (i) *pothole*, (ii) *longitudinal furrow*. In A and B, the lower drawings are cross sections along the dotted lines in the upper drawings. C: Coalesced bedforms: (i) *solution pits*, (ii) *potholes*. D: Isolate *flutes*. E: Conjugate *flutes*.

two or more centers of erosion separated by ridges (Fig. 6B). The term is applied to *concave features* or the concave parts of *composite forms*. Compound bedforms may either exhibit a hierarchical structure, containing small-scale secondary sculpted forms within a large primary feature, for example, a large *pothole* containing several smaller *potholes* in its walls and floor, or be internally subdivided into two or more *concave features* of a comparable size. The various centers of erosion or component forms may always have been present within the bedform throughout its growth; they may have been initially discrete forms that subsequently coalesced into a larger, more complex bedform; or the compound bedform may have evolved from an initially simple form, either by its subdivision through internal secondary sculpting or by its extension through external secondary sculpting. Usually the origin of compound forms is uncertain, and the term is used in a descriptive manner for bedforms with undulating or convoluted floors and walls. A special type of compound feature is the "paired" sculpted form, consisting of two similar *concave features* of comparable size with a shared edge. The two components are apparently intimately linked in their growth, development, and hydrodynamics. The outline of the two component *concave features* viewed as a whole suggests that the pair interact in a constructive manner and behave as a single, larger bedform. Another type of compound feature that lacks a hierarchical arrangement of its components is the *furrow complex* (section 1.4 of the typology), in which two or more component *furrows* with shared boundaries exhibit a distinctive geometrical arrangement. The various components within compound forms exhibit constructive interference in terms of their flow structures; this principle is discussed in a later section.

(iii) Coalescence is the process by which two or more bedforms merge to produce a single, larger bedform. During this process, boundaries between bedforms are destroyed, and coalescing individuals are subsumed within the new larger bedform, eventually losing their identity. Bedforms may be described as "coalesced" if (1) this process has proceeded to a considerable degree (even though the components, formerly separate bedforms, may remain distinguishable) and (2) an origin through coalescence may be reasonably inferred from their internal morphology and that of nearby bedforms that have not yet begun to coalesce or are in an earlier stage of the process (Fig. 6C).

Isolate and Conjugate Forms

These terms are not used in the naming of bedforms; rather, they are adjectives describing the states in which individual *concave features* may exist, and they do not appear in italics. The terms tend to lose their meaning when one considers *convex* and *undulating surfaces* because of the poorly defined lateral extent of these forms. They are used in a sense similar to that of Allen (1971a). "Isolate" bedforms are discrete and independent forms, whereas "conjugate" bedforms are those in which neighbors of the same bedform are interfering and have shared boundaries (Figs. 6D and 6E). Whereas isolate bedforms have rims separating them from the surrounding rock surface, conjugate forms have crests between contiguous individuals. "Conjugation" is the process of becoming conjugate, and it occurs through the growth and subsequent interference of initially isolate forms. In contrast to the process of coalescence, individuals are not subsumed into a single, larger feature but retain their identity within an extensive field of similar bedforms separated by crests. However, during the process of becoming conjugate, sculpted forms may experience significant changes in their character, particularly in their planform pattern. Conjugation may have progressed to varying degrees in different bedform assemblages, and the degree to which a bedform assemblage has become conjugate relates not only to what proportion of individuals are interfering and what proportion of boundaries are shared, but also to what extent the characteristics of conjugate individuals differ from those of their isolate antecedents. Thus, sculpted forms may be partially conjugate, where individuals remain similar in character to their isolate predecessors in spite of their shared boundaries, or they may be fully conjugate, where individuals appear merely as part of a larger network of contiguous bedforms, and where they are much altered in character and are no longer clearly related to their isolate predecessors. Once developed, this network of bedforms appears to become in some sense self-stabilizing and persistent, as for example in the case of *scallops*, rather than undergoing further change to some other bedform.

Sharp or Cuspate Edges

The terms "sharp" and "cuspate" as applied to the crests and rims of bedforms (in vertical section) are used synonymously in this study. Strictly speaking, "cuspate" describes a change in slope that is absolutely abrupt, i.e., is a singularity in mathematical terms. In nature, however, this is impossible, and so-called "cuspate edges" actually are curves of very small radius. Sharp edges are therefore defined here in a pragmatic way, as boundaries that appear to an observer viewing the bedforms as sharp lines. They are thus sharp on the scale of the bedforms; viewed in detail, for example, with a hand lens, they may be rounded. Edges are termed "crests" when they occur between conjugate bedforms and between the component features of a compound bedform, and "rims" when they form the boundary of isolate ones. Sharp crests may be one of three types, based on the nature of the slopes on either side:

(i) The slopes on either side are convex-up.
(ii) The slopes on either side are concave-up.
(iii) The sense of curvature of the slope changes across the crest, i.e., concave to convex.

Flutes and Scallops

The terms "flute" and "scallop" have long been used to describe the ripple-like patterns of sculpted features found in caves. However, "flute" is also used by geologists to describe the more or less parabolic, open-ended scour marks formed in cohesive substrates by turbidity currents (e.g., Dzulynski and Sanders, 1962; Dzulynski and Walton, 1965; Allen, 1965, 1968, 1969, 1971a, 1971b; Collinson and Thompson, 1982: p. 38; Leeder, 1982: p. 103), leading to some confusion. The latter group of features was named "pocket flutes" by Maxson and Campbell (1935) and simply "flutes" by Maxson (1940), who defined flutes as short, discontinuous *furrows* parallel to the flow with upstream rims that are parabolic in plan view. Unfortunately, the marks they illustrate from boulders in the Colorado River are partially conjugate, and as a result the patterns of marks have begun to develop crests with some continuity in the transverse direction (i.e., like ripples) but they nevertheless remain identifiable as flutes by virtue of the dominance of crest lines subparallel to flow. Because of this, Maxon's (1940) bedforms were apparently misinterpreted by Coleman (1949) in his paper on scour marks in caves, in which he describes fairly three-dimensional ripple-like forms, which he mistakenly identified as Maxson's (1940) flutes. However, Coleman considered "flute" to be a misnomer for his ripple-like forms, which are not *furrow*-like at all in the streamwise direction, and proposed the name "scallop" as a replacement.

Nevertheless, the terms "flute" and "scallop" have both continued in common usage in connection with caves to describe transverse, ripple-like erosion features. They became differentiated by Curl (1966), who described them as two- and three-dimensional forms, respectively, of essentially the same bedform: *flutes* have more continuous and less sinuous crests. Curl's definitions have persisted and become familiar (e.g., Goodchild and Ford, 1971; Blumberg and Curl, 1974; Allen, 1982: p. 242; Gale, 1984; Ford and Williams, 1991: p. 303–305; Slabe, 1996: p. 19).

Other terms have also been used to describe these features. Cave flutes (as defined by Curl, 1966) are regarded as identical to *solution ripples* (Allen, 1982: p. 246–247; Jennings, 1985: p. 75) by Ford and Williams (1991: p. 376, 388) and confusingly named "fluted scallops" by them. In fact, this is erroneous, since *solution ripples* are developed by sheets of water on steep or overhanging walls (Allen, 1982: p. 247), whereas flutes (sensu Curl, 1966) develop in caves on surfaces experiencing conduit flow (Gale, 1984). While a close relationship between them has been suggested on the basis of morphology (Allen, 1982: p. 247), the two should be separated until a better understanding of them is obtained. The phrases "cockles" and "cockling pattern" are also used to refer to cave flutes and scallops (Sweeting, 1972: p. 144; Ford and Williams, 1991: p. 385), but this name is more properly applied to conjugate karstic *solution pits* (Allen, 1982: p. 241). Bögli (1981) referred to scallops as "cave flow forms" or "cave facets."

In fact, the term "flute" has been applied to six environmentally and morphologically distinct groups of features:

(i) forms sculpted by subglacial meltwater (s-forms) consisting of grooves and ridges at a range of scales (Tinkler and Stenson, 1992; Pollard et al., 1996; Shaw, 1996; Kor and Cowell, 1998; Shaw et al., 2000)
(ii) the more or less parabolic scour marks developed in cohesive sediments (e.g., Dzulynski and Walton, 1965; Allen, 1968, 1971a, 1971b) and rock (Maxson and Campbell, 1935; Baker and Pickup, 1987; Hancock et al., 1998)
(iii) various downslope-orientated karstic grooves: as a synonym of rillenkarren (Jennings, 1985: p. 75; Ginés, 1996), and in the term "decantation flutings" (Ford and Williams, 1991: p. 376, 388)
(iv) various types of flow-parallel *furrows* in bedrock in fluvial environments, e.g., the "spiral flutes" of Alexander (1932) and the "sweep flutes" of Maxson and Campbell (1935), and in sand in aeolian environments (Whitney, 1983)
(v) erosional remnant rock ridges, similar in appearance to yardangs, in coastal environments (Bryant, 2001: p. 87)
(vi) ripple-like features in caves, and *solution ripples*, as described above

In view of this confusing nomenclature, it is suggested that the term "flute" be restricted to its original meaning, i.e., the more or less parabolic scour hollows of Dzulynski and Sanders (1962), Dzulynski and Walton (1965), and Allen (1968, 1971a, 1971b), and in particular that the name be dropped in connection with transverse ripple-like features in caves, which are entirely distinct in spite of the error made by Coleman (1949), and for which the name "scallop" is preferred. Thus, cave flutes (sensu Curl, 1966) are properly regarded as the two-dimensional end-member type of scallop. Hereafter in this report, *flute* shall be used in the geological sense to refer to the generally parabolic, open-ended scour marks of cohesive sediments and rock (Figs. 7A–C), and *scallop* shall be used to describe transverse ripple-like features of the type found in caves.

However, *flutes* and *scallops*, though distinct, are related. The development of *scallops* from the complete conjugation of an assemblage of initially isolate *flutes* formed at defects is well documented (Allen, 1971a, 1971b; Goodchild and Ford, 1971; Ford and Williams, 1991: p. 305–306). Once the *flutes* become fully conjugate, their character substantially changes in that the bedforms are then dominated by transverse crests rather than crests subparallel to the flow, which is why *scallops* are a distinct bedform. It is this genetic relationship which led to the misinterpretation of Coleman (1949) described above. Because Maxson and Campbell's (1935) "pocket flutes" and Maxson's (1940) "flutes" are partially conjugate, there are enough similarities between them and Coleman's cave features for him to identify them as the same bedform. To clarify this nomenclature, it is therefore suggested here that the forms of Maxson and Campbell (1935) and Maxson (1940), dominated by flow-parallel ridges and *furrows,* with parabolic crests at their upstream end but open at their downstream end, be referred to as conjugate *flutes* (Figs. 7B and 7C). Once they become fully conjugate, dominated by transverse crests, ripple-like in appearance, and no longer obviously related to their antecedent isolate *flutes*, they are named *scallops* (Fig. 7D). For example, the bedforms illustrated by Maxson and Campbell (1935: Fig. 8 therein) and described as "broad pocket flutes" are highly ripple-like and would be more correctly termed *scallops*.

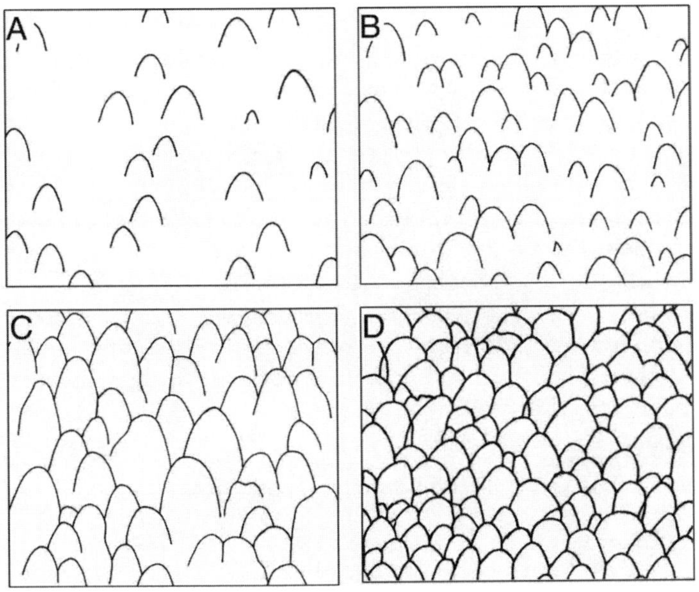

Figure 7. Cartoons illustrating the process by which a *scallop* assemblage is formed from a field of *flutes*. Initially isolate *flutes* (A) enlarge, and new *flutes* are nucleated. The *flutes* thus begin to interfere and become progressively more conjugate in nature (B, C). *Scallops* are formed when the *flutes* become fully conjugate (D).

Furrows, Grooves, Runnels, Troughs, Welts, and Channels

Curvilinear depressions have been referred to as *furrows* (e.g., Maxson and Campbell, 1935; Allen, 1965, 1971a; Collinson and Thompson, 1982: p. 40; Kor et al., 1991; Tinkler, 1993; Kor and Cowell, 1998), grooves (e.g., Maxson and Campbell, 1935; Maxson, 1940; Allen, 1969, 1971b; Shepherd and Schumm, 1974; Ikeda, 1978; Wohl, 1992b, 1993; Wohl and Ikeda, 1998; Bryant, 2001: p. 85), *runnels* (e.g., Ford and Williams, 1991: p. 376, 382–388; Hutchinson, 1996), *troughs* (e.g., Karcz, 1973; Collinson and Thompson, 1982: p. 37; Kor et al., 1991; Tinkler, 1993, 1997a; Wohl and Ikeda, 1998; Shaw et al., 2000; Bryant, 2001: p. 89), welts (e.g., Friend, 1965; Allen, 1974), and channels (Kor et al., 1991; Tinkler, 1997a) in a more or less synonymous fashion. Some standardization of this terminology is required. The following distinctions are suggested on the basis of the contexts in which they have commonly been used, and usefulness in a morphological sense.

(i) A channel is a major conduit for flow, within which the flow is confined under normal conditions. The dimensions of the flow are comparable to those of the channel. This term is therefore best avoided in descriptions of roughness elements and small-scale features.

(ii) The term *trough* is best applied to the depressions in a train of quasi-periodic, transverse bedforms such as ripples and dunes, where it has a long-accepted application and has analogy with the terminology of waves. The term has been used as a synonym of *flute* (e.g., Wohl, 1992b, 1998; Tinkler, 1997a; Hancock et al., 1998; Wohl and Ikeda, 1998). *Flute* is preferred to *trough* in the latter context, however, on the bases of precedence and established usage.

(iii) A *furrow* is a (curvi)linear depression situated within, and with dimensions much smaller than, a channel, and whose length is more than twice its width. It has relatively smooth sides and a substantially rounded cross section (although the rims may be cuspate). This is a broad definition and includes all the features that have previously been described as grooves or welts. "Groove" and "welt" are therefore considered synonyms of *furrow* and are not preferred terms. *Furrows* may follow structural controls or be unrelated to them, but the criterion that they should be rounded in cross section excludes entirely structural features such as slightly widened joint cracks, which have experienced little or no sculpting.

(iv) *Furrows* in shallow, riffle-like areas of the channel often have the effect of focusing and channeling flow at low stage if they satisfy certain criteria. These criteria include that the *furrows* occupy the lowest part of the channel, that they be approximately longitudinally orientated, and that they have a pool-like feature at their upstream ends from which water is supplied to the *furrow*. *Furrows* acting as conduits in this manner at low stage contain deeper and faster water than that present over adjacent bedrock surfaces. The adjacent bedrock may even be dry. *Furrows* channeling water in this way are termed *runnels* in this typology. At high stage, however, the *runnels* and their associated flow structures are drowned out. They no longer focus flow and instead behave with respect to the flow as any other *furrow*, i.e., as a roughness element rather than as channel topography. *Runnels* that are especially large (but still only a fraction of the width of the main channel) may be termed *inner channels*. This use of the term *runnel* is consistent with its application in karst geomorphology (Ford and Williams, 1991: p. 376, 382–388), where it refers to small-scale linear depressions acting as conduits.

Potholes

Elston (1917, 1918) and Alexander (1932) classified *potholes* according to the (inferred) responsible erosion mechanisms and flow structures respectively. However, both of these classifications are based on a broad definition of *potholes* compared with that traditionally in use at the time, and they include features that might be best described separately, such as *plunge pools*. This limitation was in fact noted by Alexander (1932). Ängeby (1951) extended the meaning of the word *pothole* further, essentially using it as a term for any type of scoured depression. A broad sense of the word appears to have become generally accepted (e.g., Blank, 1958; Allen, 1982: p. 261–264; Selby, 1985: p. 244–245; Sato et al., 1987). The effect of this is that the term has lost any scientific meaning it may once have had.

The present authors therefore adopt a stricter definition of *potholes* as essentially round (in plan view), deep depressions, which are, or can be expected to be, eroded by vortices with approximately vertical axes by mechanisms other than plucking. This is the general flow pattern assumed by many authors to be responsible for *potholes* (e.g., Alexander, 1932; Lugt, 1983: p. 144; Zen and Prestegaard, 1994). *Potholes* thus defined are equivalent to the "normal potholes" of Elston (1917, 1918) and the "eddy holes" of Alexander (1932). Features that differ significantly in shape have significantly different origins and flow structures and may be separated into other groups. This strict definition is therefore justified on both morphological and mechanical grounds. To group together many diverse bedforms under the single heading of *potholes* does not aid our understanding or description of them. For instance, the "streamside potholes" of Ängeby (1951), which are linear depressions orientated generally across the channel and which occur on the upstream sides of abrupt bedrock highs, are presumably formed by quasi-transverse attached vortices (rollers) analogous to those found in regions of flow convergence in front of *forward-facing steps* (e.g., Pollard et al., 1996). The classical *pothole*, however, is associated with vertical vorticity generated at an abrupt flow expansion at the upstream lip of the *pothole*, i.e., at a *backward-facing step*. The shallow, steep-sided circular depressions of Bryant (2001: p. 88) would also not be classified as *potholes* according to this definition.

No upper size limit is placed on *potholes*, and examples up to ~10 m in diameter are not uncommon (Alexander, 1932; Sato et al., 1987). At very large scales of bedform, the spacing of any structural weakness, and therefore also the dimensions of plucked blocks, become small compared with the dimensions of the bedform. In this case, the process of plucking becomes analogous to erosion by abrasion in that the morphologies of features produced by plucking are no longer necessarily defined by structural elements of the substrate; their shapes can reflect hydrodynamic patterns in the same way that bedforms developed through abrasion do. In principle, therefore, very large scale bedforms can be produced by plucking. An example of such a feature is the ~40 m diameter "pothole" eroded through plucking by the catastrophic Pleistocene Missoula floods described by Baker (1973). While its shallow morphology excludes this feature from being considered a true *pothole*, it is nonetheless a fluvial bedform. There must, however, be a lower size limit placed on *potholes*. At this end of the scale, *potholes* grade into *solution pits* and other small depressions in which different flow structures or mechanisms operate. The observer must assess whether the depression is of sufficient magnitude to sustain a temporally persistent and powerful, vertically orientated vortex. A diameter of ~10 cm might be the right magnitude for this lower limit.

Potholes are also sometimes referred to as "swirlholes" (Jennings, 1983; Kale and Shingade, 1987). This word has been used to avoid confusion with karstic features known colloquially by the same name. However, in view of the familiarity and popular usage of the term *pothole* in the context of fluvial sculpted forms, and arguably, its being more appropriate as a descriptive term in this sense than in the sense of the karstic features, *pothole* is retained in this report in preference to "swirlhole."

A TYPOLOGY OF BEDROCK BEDFORMS

Introduction

This section presents a proposed typology of sculpted forms in open bedrock channels, representing the results of recent fieldwork combined with a critical appreciation of published bedform descriptions. The typology is summarized in Table 1. New nomenclature is kept to a minimum, and existing or familiar

TABLE 1. SUMMARY OF THE TYPOLOGY OF SCULPTED FORMS, SHOWING THE FIRST THREE SUBDIVISIONS

1. Concave features

1.1 Potholes
- Simple potholes
- Potholes with external furrows
- Compound potholes
- Breached potholes
- Lateral potholes

1.2 Longitudinal furrows
- Flutes
- Short furrows
- Parallel-sided furrows
- Compound parallel-sided furrows
- Expanding furrows
- Runnels with wall relief

1.3 Nonlongitudinal furrows
- Oblique sloping furrows
- Transverse furrows
- Reversed furrows

1.4 Furrow complexes
- Convergent
- Yin-yang
- Bifurcating
- Nested-curved

1.5 Overhanging concave features
- Overhanging concave surface
- Cavetto, taffoni, alcove

1.6 Shallow concave surfaces
- (Not subdivided)

2. Convex and undulating surfaces

2.1 Hummocky forms
- Nondirectional hummocky forms
- Longitudinal hummocky forms
- Transverse hummocky forms
- Sharp-crested hummocky forms

2.2 Other convex and undulating surfaces
- Partially abraded surface, polished surface
- Upstream-facing convex surface
- Irregular sculpting, convoluted surface
- Bladed forms, faceted obstacle

3. Composite forms

3.1 Obstacle marks
- Simple obstacle marks
- Compound obstacle marks

3.2 Hummocky forms with linear depressions
- Pseudoripples with longitudinal furrows
- Runnels with bed relief
- High-relief hummocky forms with transverse furrows

3.3 Convex surfaces with steep lee faces

Hummocky form with steep lee face, streamlined steep lee face

4. Solutional forms

4.1 Solution pits and pans
- Fluvial solution pits
- Nonfluvial solution pits and pans

4.2 Scallops
- Nondirectional scallops
- Directional scallops

4.3 Other Solutional Forms
- Vertical ridges, solution ripples, decantation runnels, longitudinal ridges, microrills, taffoni

5. Tool marks

5.1 Percussion marks
- Edge percussion marks
- Face percussion marks
- Cavitation percussion marks

5.2 Scratch marks
- Simple scratch marks
- Looped and arcuate scratch marks
- Parallel scratch marks

6. Large-scale sculpted features

6.1 Uniform bed gradient
- Discordant plane bed
- Inner channel
- Out-of-phase undulating walls
- In-phase undulating walls
- Yardang

6.2 Variable bed gradient

Knickpoint, knickzone
Cataract
Step
Chute
Bifurcating chute
Plunge pool, gouge hole
Step-pool system
Pool-riffle/pool-rapid system
Breached pools
Convex surface
Incised convex surface

names are conserved wherever possible. This typology is intended to clarify, revise, and add to existing descriptions. It is as complete as possible at present but is unlikely to be comprehensive, due to the wide variety of fluvial forms. Nevertheless, new bedforms identified in the future should be easily accommodated within the framework of this typology. In spite of the variety of fluvial forms, and while each reach possesses a character of its own in terms of the bedforms present and their style, a degree of universality of bedforms is observed, whereby features common to many channels may be identified, at some level of similarity. This typology contains four levels of subdivision and identifies around 140 bedform types. At the third level of subdivision, representatives of each of the most important groups, such as *simple potholes*, *flutes*, and *short furrows* will be found in the majority of bedrock channels containing sculpted forms.

Many field sites, exhibiting a range of rock types, were visited during fieldwork undertaken for this study. The bedforms observed at all of these sites contributed toward the development of the ideas and typology presented. Thirty-four sites are illustrated by photographs in this report, thirty-two of which are rivers. These river locations are summarized in Table 2, which also gives lithologies. Full details of the sites would add little useful information and for the sake of brevity are not given. Local site names are given where this is useful, for instance, where there is more than one site illustrated on the same river. The reaches examined vary considerably in length from a few tens of meters to several kilometers. The rivers include perennial streams from temperate and tropical regions, and ephemeral streams from monsoonal and hyperarid climates.

A common method of classification of alluvial channel features is into macroscale, mesoscale, and microscale forms. Sedimentary bedforms in these categories scale with channel width, flow depth, and particle size respectively (Best, 1996). A similar scheme has also been applied to bedrock channels (Baker, 1978; Wohl, 1998; Wende, 1999). It is not adopted here, however, because bedrock and alluvial bedforms are not directly analogous in several respects, as noted above, and because the controls on their scaling are as yet unclear (see Discussion for a list of factors that may control the scale of bedrock bedforms). To extend the analogy between bedrock and alluvial bedforms in this way is therefore misleading.

The classification used in this report is based mainly on bedform morphology and orientation, rather than on inferred processes or other characteristics. This is partly because current understanding of the processes by which these features are formed and their associated flow structures is slim, and therefore any inferences may be incorrect, but also because a morphological classification is of greatest use for field identification and for comparison with published accounts. Furthermore, in view of the assumed close link between morphology and flow structure, it is thought that the bedform groups, although based on morphology and orientation, nevertheless also reflect major differences in the associated flow structures (see, e.g., Springer and Wohl's [2002] analysis of morphology and flow fields in cave dissolution features), and these are discussed in the descriptions that follow. However, inferred processes and flow characteristics are used as classifying characteristics where necessary to avoid grouping together forms with broadly similar morphologies, but which occur in widely differing hydrodynamic environments or which develop through different inferred processes. All the features described below are hydrodynamically controlled erosional forms, unless stated otherwise.

The typology presented here focuses on relatively small-scale forms, i.e., those whose dimensions are much less than the channel width. In practice, this corresponds to the millimeter to meter scale. Bedforms in this size range both affect and respond to very local flow patterns. Three major groups of these forms are identified, based on topology:

(i) *concave features*: various types of *potholes*, *furrows*, and other miscellaneous *concave features*

(ii) *convex* and *undulating surfaces*: *hummocky forms* and other miscellaneous *convex* and *undulating surfaces*

(iii) *composite forms*: bedforms such as *obstacle marks* and *hummocky forms with linear depressions*, comprising elements from both of the above groups

Two further groups of small-scale forms are defined based on inferred processes:

(iv) *solutional forms*: *solution pits* and *pans*, *scallops*, and other *solutional forms*

(v) *tool marks*: *percussion marks* and *scratch marks*

A final section considers larger-scale fluvial forms, those of dimensions up to that of the channel width or larger, which interact with the main body of the flow and generally act as topography. These are described mainly from existing literature because these features have not been the main focus of fieldwork undertaken for this report, although some new types are introduced. The s-forms, sculpted forms of subglacial origin, will not be described, although they are referred to. Good descriptions of them may be found in Kor et al. (1991), Shaw (1996), and Kor and Cowell (1998).

Within this study, reference will often be made to the local flow direction, which may vary greatly from the general orientation of the channel as a result of the complex flow that exists in bedrock channels, frequently even being directed upstream. On surfaces exposed at low stage, and on which bedforms are observed, the direction of local flow at high stage when the surfaces are submerged and the bedforms active cannot be assumed and must be carefully assessed, based in the first instance on the local channel topography. This includes observing whether there are any channel features, such as bends, hollows, projections, and boulders, which might perturb the flow. If there are no such features, local flow may be assumed to be parallel to the general direction of the channel. Subsequently, once some knowledge of the orientation of different types of sculpted forms with respect to local flow has been acquired, this assessment of local flow direction combines local topography with the observation of the distribution, type, and orientation of bedforms. The interpretation of flow structures within sculpted forms and the use of sculpted forms as indicators of flow direction

TABLE 2. BRIEF DETAILS OF RIVERS ILLUSTRATED IN THIS REPORT, IN THE ORDER IN WHICH THEY APPEAR

River name	Country	Region	Site name (if any)	Lithology
River Stinchar	United Kingdom	Southwest Scotland		Mudstone
River Eden	United Kingdom	Northwest England		Limestone
River Dee	United Kingdom	Cumbria, Northwest England		Limestone
Sichiri-gawa	Japan	Chiba Prefecture		Interbedded sandstone and mudstone
Borrow Beck	United Kingdom	Northwest England		Calcareous mudstone
River Lune	United Kingdom	Northwest England	Tebay	Calcareous mudstone
Nam Mae Chaem	Thailand	Northwest Thailand	Ob Luang	Granitic gneiss
River Lune	United Kingdom	Northwest England	Halton	Fine-grained sandstone
Woolshed Creek	Australia	Northern Victoria	Woolshed Falls	Granite, microgranite
Nam Mae Khlang	Thailand	Northwest Thailand		Gneiss
Nam Mae Chaem	Thailand	Northwest Thailand	Ob Noi	Granitic gneiss
Wadi Hazazon	Israel			Limestone, dolomite
Huai Nang Rong	Thailand	East-central Thailand		Rhyolitic agglomerate
Watson River	Greenland	Western Greenland	Kangerlussuaq	Gneiss
Birk Beck	United Kingdom	Northwest England		Medium-grained sandstone
Nahal Zin	Israel		Lower Nahal Zin	Limestone
Nahal Hatira	Israel		Ein Yorkeam	Interbedded limestone and marl
Than Rattana	Thailand	East-central Thailand		Andesite
Nahal Zin	Israel		Ein Avedat	Limestone
Stakeley Beck	United Kingdom	Northwest England		Medium-grained sandstone
Nahal Ashalim	Israel			Limestone
Nahal Shani	Israel		Hakenyon Ha'adom	Fine-grained sandstone
Nahal Neqarot	Israel			Limestone
Allt Ceitlein	United Kingdom	Northwest Scotland		Granite
Gayle Beck	United Kingdom	Northwest England		Limestone
Som Ploy	Thailand		Som Ploy Waterfall	Fine-grained sandstone
River Wharfe	United Kingdom	Northwest England		Limestone
Katun River	Russia	Southern Siberia		Granite
Sleightholme Beck	United Kingdom	Northeast England		Limestone
River Kent	United Kingdom	Northwest England		Limestone
Ease Gill	United Kingdom	Northwest England		Limestone
River Severn	United Kingdom	Southwest England		Mudstone
Nishi-zawa	Japan	Yamanashi Prefecture		Granite

are discussed in a later section. However, it is not always possible to predict in this manner the local flow direction over a surface when submerged. In the illustrations that accompany this typology, the local flow direction is given where it is known, for example, "Flow from left to right." Where it is not known, the upstream direction for the channel as a whole is given, but this should not necessarily be regarded as the direction of flow over the bedform.

An important point should be made before proceeding to the descriptions of bedforms. In the field observations made for this report, perhaps the main underlying principle of sculpted features in bedrock channels was found to be the continuity of form between end-member types. The researcher instinctively attempts to identify a limited number of ideal types of forms that are easily recognizable and repeated from place to place, even though examples may deviate from the ideal type. Although such ideal types do exist, what is most striking is not the relatively small amount of variation about a limited number of readily identifiable type forms, but the wide and continuous spectrum of forms between such ideal types, by which one bedform type merges with another. Thus, most of the bedforms one can define are best considered as end members of a continuous spectrum of forms, or as midpoints within it, rather than as discrete types of forms. The idea of a continuum between bedforms in terms of form and associated flow structures was hinted at by Alexander (1932) in his discussion of the range of bedforms he described as "potholes." Such a situation was also found to exist in the experiments of Allen (1971a), in which parabolic *flutes*, *spindle-shaped flutes*, and *furrows* were found to be closely related, with erosional marks evolving between these states with time. Furthermore, individual sculpted features are often complex and may possess some of the characteristics of several apparently disparate "ideal" bedform types. Both of these facts illustrate that although there may at first appear to be a bewildering array of sculpted forms, many forms are in fact closely related. The idea of continuity is discussed further in a later section, but it is mentioned here because it is used repeatedly in the typology presented below, and should be constantly borne in mind when considering the bedform descriptions.

1. Concave Features

This group contains a wide spectrum of *potholes* and *furrows* under various subheadings. These bedforms represent highly localized regions of enhanced scour, creating a concavity. The group essentially consists of a spectrum of depressions exhibiting wide and continuous variation in the degree of elongation. It includes at one end of the spectrum *circular potholes* with no elongation at all, and at the other, highly linear *furrows*, without discrete breaks in this sequence. *Furrows* within the group also exhibit a range of orientations, from *longitudinal* (parallel to flow) through *transverse*, to *furrows* carrying reversed flow.

1.1 Potholes

These are essentially deep, generally closed depressions, which are round to elliptical in plan view, although there are many varieties (Fig. 8). They range from saucer-like depressions to shafts many times deeper than wide. The role of grindstones, once held to be the main agent of *pothole* formation (e.g., Elston, 1918; Ives, 1948), is now doubted by many, and abrasion by suspended load is generally considered the most important mechanism (Alexander, 1932; Maxson and Campbell, 1935; Allen, 1982: p. 261, 288–289; Wohl, 1992b, 1993; Zen and Prestegaard, 1994; Wohl and Ikeda, 1997; Whipple et al., 2000a). They contain vortices with near-vertical axes in general (Alexander, 1932; Lugt, 1983), although *pothole* axes may deviate significantly from the vertical in any direction. Alexander (1932) demonstrated experimentally how this may come about as a result of the corkscrew nature of the axes of vortices in cylinders of water. Ängeby (1951), Sato et al. (1987), and Wohl (1992b) illustrate *potholes* with inclined axes. Kunert and Coniglio (2002) describe *potholes* with inclined and/or sinuous axes. *Potholes* develop apparently at the site of some preexisting bed defect that generates vertical vorticity and that either evolves into a *pothole* or has one or more *potholes* scoured within it (Elston, 1918; Nemec et al., 1982; Jennings, 1983; Lorenc et al., 1994).

As a working and arbitrary definition, it may be said that the depth of a *pothole* is at least a quarter of the width, and that *potholes* have a length not more than twice their width or, in the case of more complex forms such as a *pothole with exit furrow*, contain within them a near-circular depression with a length not more than twice its width. The walls must also be vertical or near-vertical close to the rim. The classification below does not primarily consider change of shape or diameter

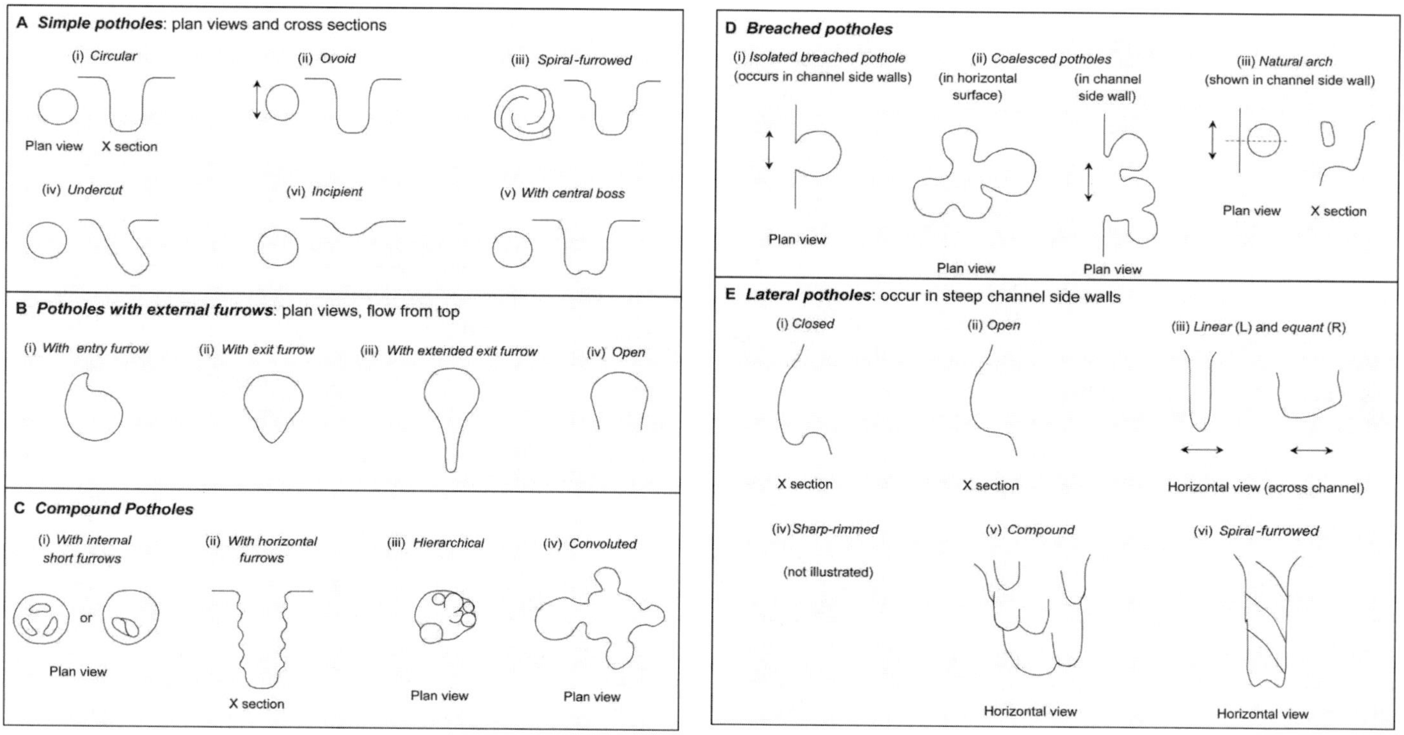

Figure 8. *Potholes*. Arrows indicate flow directions. Dashed lines (where present) indicate lines of cross sections.

with depth in the hole, although such variations are probably the norm (Ives, 1948; Ängeby, 1951; Sato et al., 1987; Lorenc et al., 1994). In some cases, the orientation of the long axis also varies in a consistent manner with depth within the *pothole* (Sato et al., 1987). These characteristics of *potholes* are often difficult to assess in the field without detailed study. Rather, the classification considers mainly the planform of the *pothole* rim and the detailed internal morphology of the *pothole.* Various combinations of the characteristics listed below can and do occur in a single *pothole*. Nemec et al. (1982) and Lorenc et al. (1994) suggested that in some areas with a range of *pothole* types, the different forms are linked together in an evolutionary sequence. *Potholes* are also known to occur in subglacial environments (e.g., Gilbert, 1906; Gjessing, 1967; Johnsson, 1988), where they are formed by meltwater, and are included as a type of s-form (Kor et al., 1991; Shaw, 1996; Kor and Cowell, 1998). However, it should be remembered that features resembling *potholes* can be formed through subaerial weathering (Johnsson, 1988). *Potholes* are often described from the geological record, both in bedrock substrates (i.e., in unconformities, e.g., Lugn, 1939; Shrock, 1948: p. 230–236; Diffendal, 1982; De Wit, 1999; Spaggiari et al., 1999; Jacob et al., 1999), and formed contemporaneously in sedimentary environments (e.g., Dixon, 1921: p. 29; Allen, 1974, 1982: p. 263–264). Kale and Shingade (1987) carried out a morphometric analysis of *pothole* dimensions in the Indrayani River, India, and concluded that *pothole* length and depth in their study area follow a power law relationship.

(a) *Simple potholes.* This group contains uncomplicated features that lack internal secondary sculpting and that do not exhibit paired relationships. The (locally) upstream rims are often sharp, but the downstream rims display varying amounts of erosion and rounding.

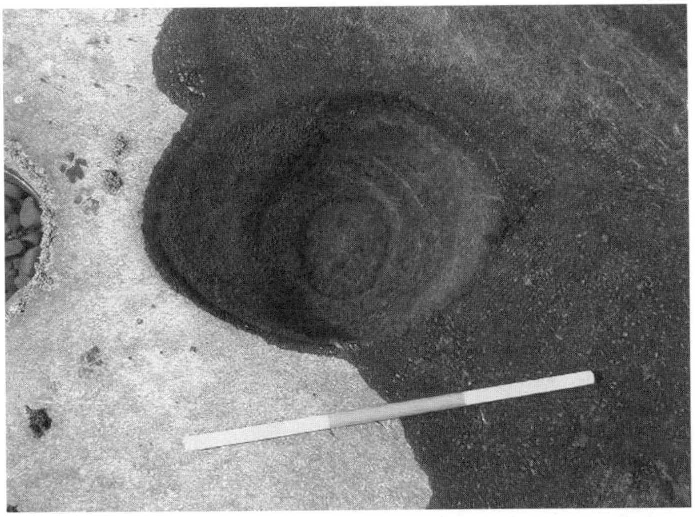

Figure 9. *Ovoid pothole*. This *pothole* has a length:width ratio of 1.35. River Lune (Halton), UK. Flow from right to left. Fine-grained sandstone. The scale bar is 60 cm long.

(i) *Circular pothole*: a depression with a circular or near-circular rim.

(ii) *Ovoid pothole*: elongated in the direction of current flow (with a length up to twice the width) (Fig. 9). Sato et al. (1987) measured the ellipticity of horizontal sections of potholes in the Kurokawa River, Japan, and showed that they have a bimodal distribution with peaks at zero (i.e., circular) and 0.3 (slightly elliptical). They also showed that the ellipticity of horizontal sections varies with depth within a hole, and that the long axis of the horizontal sections of *potholes* commonly rotates in a systematic way with depth. *Ovoid potholes* are also described by Nemec et al. (1982) and Kunert and Coniglio (2002).

(iii) *Spiral-furrowed pothole*: potholes with *spiral furrows* ornamenting the walls have been frequently described (e.g., Alexander, 1932; Maxson and Campbell, 1935; Ängeby, 1951: Figs. 23 and 24 therein; Gjessing, 1967; Morgan, 1970; Allen, 1982: p. 262; Diffendal, 1982; Jennings, 1983; Baker and Pickup, 1987; Kor et al., 1991; Wohl, 1992b; Shaw, 1996; Glasser and Nicholson, 1998; Kor and Cowell, 1998) and are generally ascribed to the presence of downward-spiraling flow thought to be present around the circumference of *potholes* (Alexander, 1932; Lugt, 1983: p. 144). Two varieties of *spiral furrows* can be identified: those found as distinct but shallow *furrows* in an otherwise smooth, rounded wall (Fig. 21), and

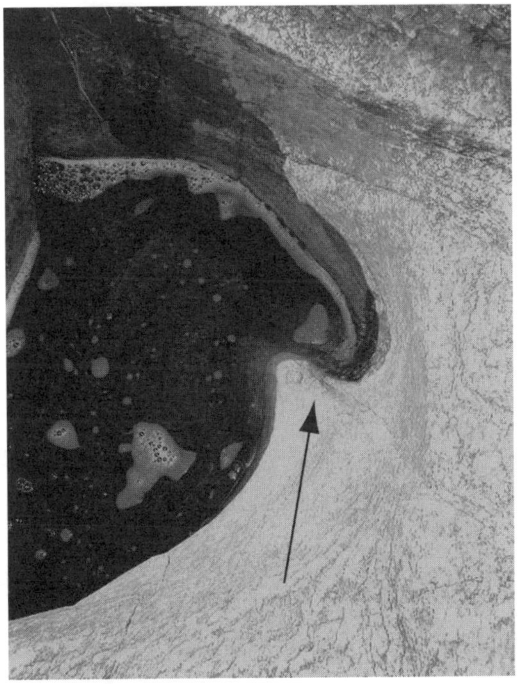

Figure 10. *Spiral-furrowed pothole* of the variety with deep incision, generating a spiral rib (arrowed) adjacent to the *furrow*. Woolshed Creek, Australia. Microgranite. The *pothole* is ~1.5 m across in its short dimension.

those which are incised deeply such that they create an adjacent projecting spiral ridge in the wall (e.g., Ängeby, 1951: Fig. 23 therein; Gjessing, 1967: Fig. 5 therein) (Fig. 10). *Spiral furrows* occur only in deep *potholes*. *Potholes with spiral furrows* are considered to be simple rather than compound forms because the internal sculpting reflects the fundamental flow structure within the *pothole* and is therefore a primary feature.

(iv) *Undercut pothole*: In some *potholes* the diameter first increases and then decreases with depth in the hole, producing a peak width at some distance below the rim; such *potholes* are *undercut* (e.g., Alexander, 1932; Ives, 1948; Nemec et al., 1982; Lorenc et al., 1994; Kunert and Coniglio, 2002). Both symmetrical and asymmetrical styles of undercutting have been identified (Nemec et al., 1982; Lorenc et al., 1994). In the asymmetrical type, the undercutting is greatest on the downstream side. A *pothole* observed by one of the authors increased in diameter with depth, then decreased forming a "waist," before increasing in diameter again.

(v) *Pothole with central boss*: This has a central high point in the floor (Fig. 11). Ives (1948), Morgan (1970), Allen (1982: p. 262–263), Nemec et al. (1982), Jennings (1983), Sato et al. (1987), Lorenc et al. (1994), and Wohl and Ikeda (1997) also describe such *potholes*. A *central boss* within a *pothole* can be readily explained with reference to the cyclonic flow pattern thought to exist within *potholes* (Alexander, 1932; Lugt, 1983: p. 144). The floor at the center of the hole is a site of low shear stress and low fluid velocities, where particles present in the fluid tend to be deposited, and would therefore be expected to be a site of reduced scour.

(vi) *Incipient pothole*: Although these are shallow (depth up to a quarter of their width), smooth, saucer-shaped depressions lacking vertical walls (Fig. 12), they are common in association with fully developed *potholes*. This fact, and their near-circular shape, suggest that they are the early stages of *pothole* development. These are the "type A" *potholes* of Nemec et al. (1982: Fig. 3 therein) and Lorenc et al. (1994: Figs. 4C and 5B therein), who also thought them to be immature *potholes*. In fact, field observations show that in many sites where *potholes* occur, there is a continuum in size and form between (1) well-developed *potholes*, (2) the saucer-shaped depressions described above and various smaller, shallower depressions such as *solution pits*, *flutes*, and *short furrows* (see below for descriptions of these forms), and (3) subangular hollows resulting from the plucking of blocks. This observation suggests that *potholes* will develop from any preexisting defect that attains sufficient size to generate powerful, vertically orientated vortices. This mode of origin was also suggested by Elston (1918), Nemec et al. (1982), and Lorenc et al. (1994). *Incipient lateral potholes* (see section e below for description of *lateral potholes*) logically must also occur, but distinguishing them from among the range of shallow concave features that typically is found on vertical and near-vertical sculpted rock faces may not be possible in general. *Incipient lateral potholes* are, however, described as a distinct bedform type from a

Figure 11. *Pothole with central boss*. Note also the *horizontal furrows*. Nam Mae Chaem (Ob Luang), Thailand. Upstream to left. Granitic gneiss. Coin on top of boss for scale.

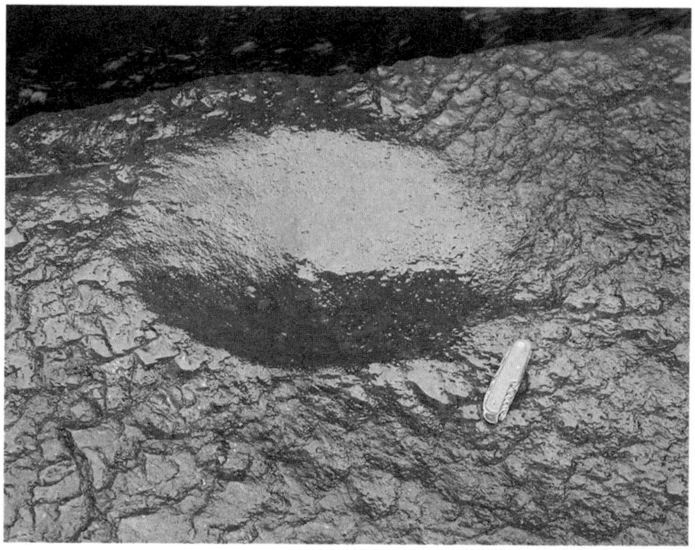

Figure 12. *Incipient pothole*. Sleightholme Beck, UK. Flow from bottom right to top left. Limestone. Penknife for scale.

cave setting by Springer and Wohl (2002). These authors refer to the subspherical bedforms as "pockets" and infer that they originate at defects in the rock or at overdeepened *scallops*. Based on their observations, Springer and Wohl (2002) develop a geometrical model of the growth of sculptures described by truncated spheres, relating the effect of bedform size and retreat rate of the surrounding surface to the necessary relative erosion efficiencies and excesses required for sculpture survival and growth.

(b) *Potholes with external furrow(s).* *Potholes* exhibit a varying degree of wear at one or more points on the lip, especially on the downstream side (with respect to the local flow pattern). In cases in which this wear is significant and results in a *furrow* that forms part of the rim and extends away from the rim, the *pothole* is said to have either an *entry* or *exit furrow*, depending on where the *furrow* occurs in relation to the hydrodynamic pattern of the bedform. These *furrows* indicate the course of powerful, erosive currents associated with the *pothole*. *Entry* and *exit furrows* are not always on the upstream and downstream sides respectively of *potholes* with regard to the flow direction in the main channel, because of the complex flow patterns that exist in bedrock channels, and because of the numerous flow recirculation cells.

(i) *Pothole with entry furrow(s)*: This has one or more slightly curved *furrows* entering the *pothole* more or less tangentially. Figure 13 shows an *entry furrow* that is unusually large in relation to the *pothole*; Figure 18 illustrates an *entry furrow* of more usual proportions. The *pothole* in Figure 168 is interpreted as having two *entry furrows* (see later section on bedforms as indicators of flow patterns). *Entry furrows* are essentially large, deep *spiral furrows* in the topmost part of the *pothole*, and they extend outside the rim of the *pothole* (e.g., Sato et al., 1987). They feed into *spiral furrows*, where present, lower down in the *pothole*, and are apparently part of the same system. *Entry furrows* can, however, exist in the absence of *spiral furrows* within the main body of the *pothole* and are thus considered separately. Like *spiral furrows*, they are evidence of the flow pattern present when the *pothole* is active, and they indicate the position of downward-directed, tangential jet(s), similar to those postulated by Alexander (1932), entering the *pothole*. Allen (1982: p. 264) describes *potholes with entry furrows,* while Gjessing (1967: Fig. 2 therein) illustrates a *pothole* with an *entry furrow* at its far right-hand side.

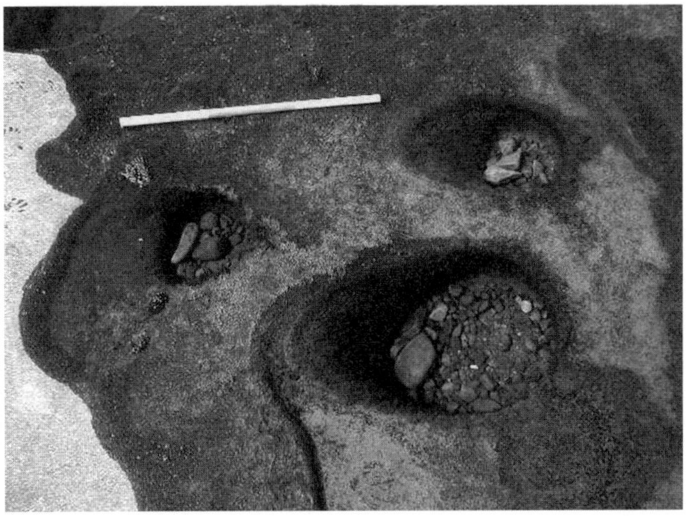

Figure 14. *Potholes with exit furrows* (seen through water). River Lune (Halton), UK. Flow from right to left. Fine-grained sandstone. The scale is 60 cm long.

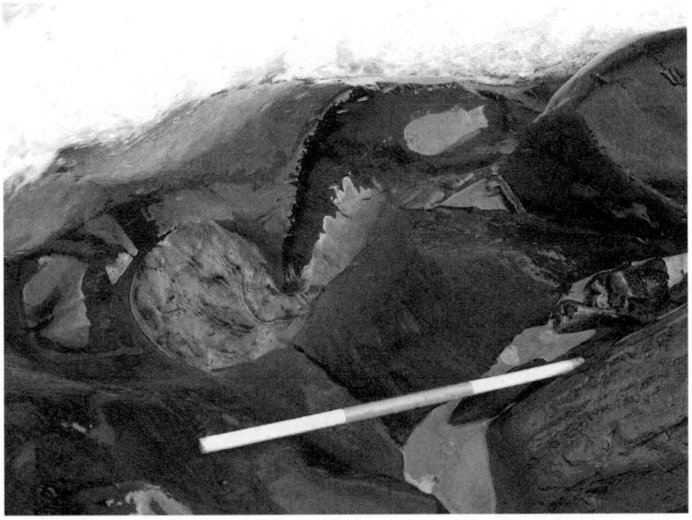

Figure 13. *Pothole with entry furrow*. River Lune (Tebay), UK. Upstream to left, but flow is apparently locally reversed at high stage, when the *pothole* is active. Calcareous mudstone. The scale is 60 cm long.

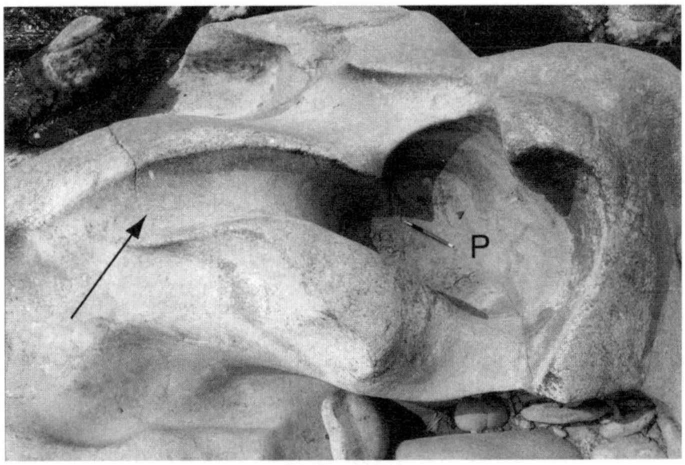

Figure 15. *Pothole with extended exit furrow*. This is a *compound pothole* (P) with several centers of erosion. The *extended exit furrow* is arrowed. Nam Mae Chaem (Ob Luang), Thailand. Flow from top right to bottom left. Granitic gneiss. Pen for scale.

(ii) *Pothole with exit furrow*: This has a straight, radially orientated *furrow* on the (locally) downstream side worn by water exiting the *pothole* (e.g., Morgan, 1970; Allen, 1982: p. 264; Kor and Cowell, 1998) (Figs. 14 and 168). The *exit furrow* is usually broad with noncuspate sides. Its depth may only be a small fraction of that of the *pothole*, or it may be a significant fraction of it. Water is thought to exit from the center of the aperture of the *pothole* after rising up its axis (Alexander, 1932; Lugt, 1983: p. 144). Tangential *exit furrows* are therefore thought by the authors not to exist on theoretical grounds. Kor and Cowell (1998) described *potholes* with both *entry* and *exit furrows*.

(iii) *Pothole with extended exit furrow*: This has a long (greater than the width of the *pothole*), well-developed *exit furrow* (Fig. 15), which may be curved.

(iv) *Open pothole*: Where the *exit furrow* becomes comparable in width and depth to the *pothole* and is relatively long, the *pothole* may be said to be open on the lee side (Fig. 16). *Open potholes* are distinguished from *breached potholes* (section 1.1d).

(c) *Compound potholes.* These are *potholes* with secondary sculpting, either on a small scale (compared with the dimensions of the primary *pothole*) internally, or in the manner of paired forms. As described by Nemec et al. (1982) and Lorenc et al. (1994), secondary sculpting is associated with the larger and more mature *potholes*. The presence of secondary sculptures within a primary *pothole* is indicative of (1) the development of small-scale vortices at the floor or walls of the primary bedform and (2) the breakup or subdivision of a single, large, coherent flow structure at those points on the floor or walls of the primary bedform.

(i) *Pothole with curved short furrows*: Many *potholes* have *curved short furrows* much smaller in width than the *pothole*'s diameter, sculpted into their floors (e.g., Gjessing, 1967: Fig. 5 therein; Morgan, 1970; Sato et al., 1987: Fig. 46 therein; Whipple et al., 2000a). The *short furrows* may have an asymmetrical sharp crest between them (Gjessing, 1967), and they may be arranged in series in an approximately circular plan, or they may be in more complex patterns such as the *yin-yang furrows* of Whipple et al. (2000a). The "yin-yang" relationship is described below as a type of *furrow complex* (section 1.4b). *Yin-yang furrow complexes* within *potholes* indicate the sense of rotation of fluid (see Discussion). Where *short furrows* are present within a *spiral-furrowed pothole*, they are present at the lower ends of the *spiral furrows* (Gjessing, 1967).

(ii) *Pothole with horizontal furrows*: The *horizontal furrows* adorning the walls of the *pothole* may have rounded rims and be without definite ends (Fig. 11), or they may have cuspate rims and distinct ends (Fig. 17). The *horizontal furrows* are regarded as secondary sculpted forms. Nemec et al. (1982) and Lorenc et al. (1994) described *horizontal furrows* as typical of larger and more mature *potholes*. Gjessing (1967) describes *potholes* with both *horizontal furrows* and *spiral furrows*.

(iii) *Hierarchical pothole*: a large primary *pothole* with one or more smaller, secondary *potholes* within it, either in the floor or cut into and perched within the sidewalls (Fig. 18). Those in the sidewalls are usually *lateral potholes* (section 1.1e). Secondary *potholes* appear to form in many cases from enlarged *spiral furrows* and from *entry* and *exit furrows* that develop irregularities. As they grow, the secondary *potholes* themselves may acquire these features, ultimately leading to a third generation of sculpting (i.e., tertiary sculpting). Nemec et al. (1982), Lorenc et al.

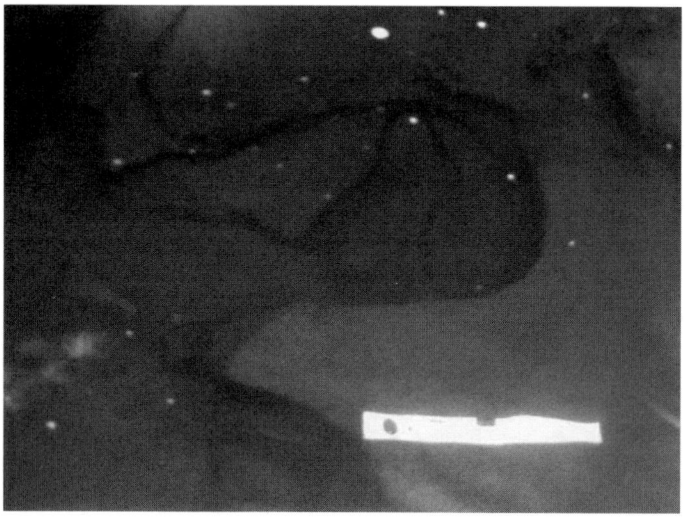

Figure 16. *Open pothole*. River Lune (Halton), UK. Flow from right to left. Fine-grained sandstone. The scale is 60 cm long.

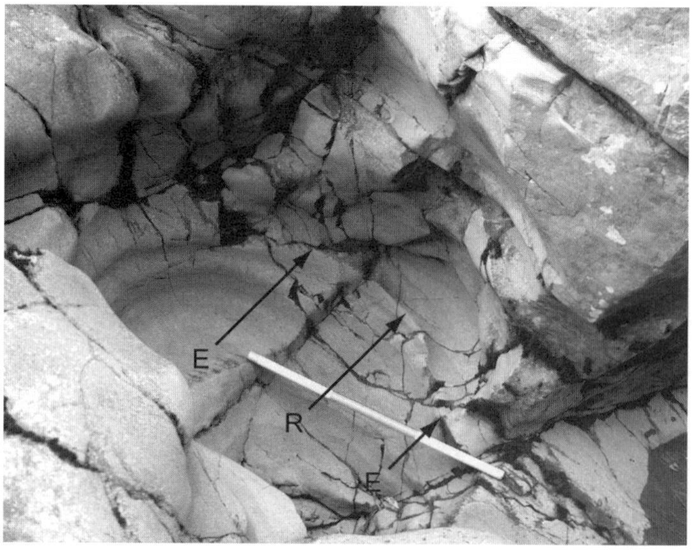

Figure 17. Interior of a *pothole with horizontal furrows*. Note the cuspate rim (R) of the *furrow* and its distinct ends (E). River Lune (Tebay), UK. Upstream to right. Calcareous mudstone. The scale is 60 cm long.

(1994), and Whipple et al. (2000a) also describe *hierarchical potholes*. Kale and Joshi (2004: Fig. 3 therein) illustrate *potholes* developing in artificial bedrock pits; these are effectively secondary *potholes* in which the primary feature is man-made. The lower part of the filled *pothole* of the basal Gering Formation (?Miocene) of Nebraska described by Diffendal (1982) appears to exhibit a hierarchical structure. The nested arrangement of the component forms allows *hierarchical potholes* to be easily distinguished from *coalesced potholes*, in which neighbors are of similar magnitude and do not have a nested or perched relationship.

(iv) *Convoluted pothole*: a catchall term for various types of *compound potholes* that possess a complex or irregular plan view and/or cross section. These range from features consisting of *potholes* paired with *short furrows*, *flutes*, or a second *pothole* (Fig. 19), to those which contain such a degree of internal secondary sculpting of various kinds that the morphology of the *pothole* has become irregular and chaotic. Jennings (1983) refers to complex *pothole* forms from the Snowy River, New South Wales, resembling "writhing, living things."

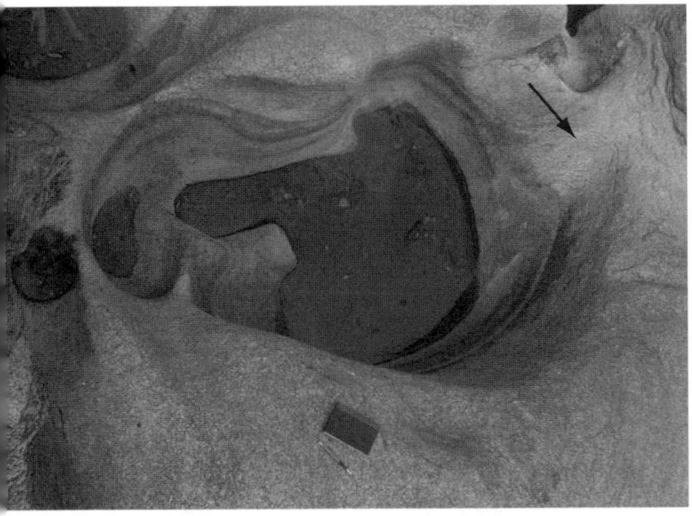

Figure 18. *Hierarchical pothole*. In addition to a complex *compound* morphology, this *pothole* has an *entry furrow* (arrowed). Nam Mae Chaem (Ob Luang), Thailand. Upstream at top of photograph. Granitic gneiss. The notebook is 15 cm long.

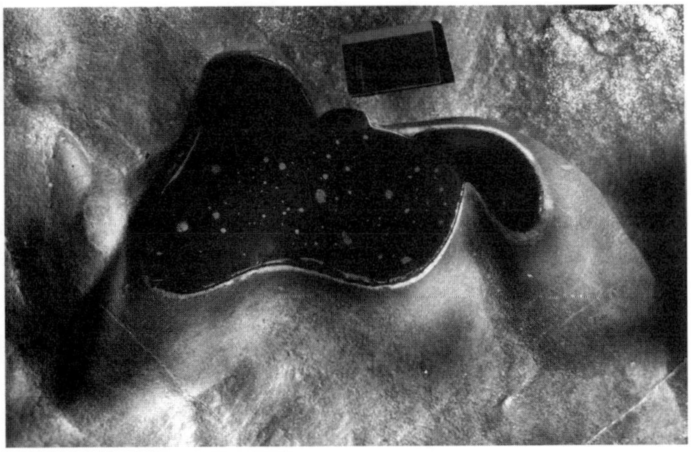

Figure 19. *Convoluted pothole*. Nam Mae Khlang, Thailand. Flow from top to bottom. Gneiss. The notebook is 15 cm long.

(d) *Breached potholes*. This is a family of *potholes* that, although once intact, have since experienced the breaching of their wall or floor through intense and prolonged abrasion, or through the plucking away of part of the wall. In this manner, a single *pothole* in a steep channel margin or close to an *inner channel* becomes united with the main flow, or two or more *potholes* coalesce. These forms are distinguished from *open potholes* (section 1.1b), which can occur on any surface, which are not necessarily united with any channel or other *pothole*, and which arise specifically through the wearing away of the downstream rim. *Breached potholes* are also distinguished from *lateral potholes* (section 1.1e), which have been inferred on morphological grounds by Zen and Prestegaard (1994) never to have been fully enclosed (see below). In contrast, no guidelines exist for the recognition of *breached potholes*, but the following are suggested:

(i) *Breached potholes* will tend to have an internal morphology characteristic of complete *potholes* (i.e., essentially a shaft, being approximately cylindrical, though not necessarily either vertical, of uniform diameter or perfectly circular in cross section), which is unrelated to the morphology of the surrounding rock surface and which gives the appearance of a *pothole* that has been sliced open in a vertical plane (Fig. 21). *Lateral potholes*, however, are not essentially open shafts and form a group with an internal morphology distinct from complete *potholes*. Most *lateral potholes* have rounded rims and an internal morphology that is continuous with, and that is indicative of development in conjunction with, the surrounding rock face (Zen and Prestegaard, 1994).

(ii) Recognition of the breaching of a *pothole* depends greatly on how much of the rock surrounding the *pothole* has been removed. If the breaching is recent, then the amount of wall removed is small, and the origin of the feature is obvious. If large amounts of the wall have been removed, then the relief of the feature will be reduced and its interior and edges potentially resculpted, so that it is impossible to recognize. The nearer the walls of the feature are to enclosing 360° of a circle, the more likely it is to be a *breached pothole*. In contrast, some of the *lateral potholes* illustrated by Zen and Prestegaard (1994) represented only small fractions (less than 20% in angular terms) of a complete circle, and the relief (measured horizontally from the plane of the vertical rims to the back wall) of such features is a much smaller fraction still of the diameter of that circle. If a *breached pothole* were being gradually removed by erosion of the surrounding surface, it seems unlikely that it would be recognizable in any distinct form once the erosion had progressed to this degree. Thus, distinct concave features enclosing small fractions of a full circle are unlikely to be *breached potholes*.

(iii) The presence of a relict plug of sediment exposed at the bottom of the *pothole*, attached to the remaining walls, but no longer fully supported, is evidence that the *pothole* has been breached (Fig. 20).

(iv) The presence of a *natural arch* or *natural pillar* (see below) is also evidence that a *pothole* has been breached.

The following types of *breached potholes* may be recognized:

(i) *Isolated breached pothole*: a shaft drilled into steeply sloping channel margins or into a *bedrock bench* adjacent to an *inner channel*, whose wall and/or floor has been eroded through (breached) in one place, thus joining the interior of the *pothole* with the main flow (Fig. 21). Ives (1948) describes and illustrates an *isolated breached pothole*, while Lorenc et al. (1994) illustrate several. Sato et al. (1987) referred to *breached potholes* as "half" or "broken" *potholes*. The features described by Whipple et al. (2000a) as "half potholes" may also be *breached potholes*. However, the features described as "breached potholes" by Wohl and Ikeda (1997) from flume experiments with a cohesive substrate would more properly be called *lateral potholes* since they were seen to evolve through the enlargement by scour of small sidewall hollows and were never fully enclosed.

(ii) *Coalesced potholes*: These consist of two or more adjacent potholes, the dividing walls of which have been breached. This would include the "twinned," "composite," and "heart-shaped" *potholes* of Sato et al. (1987). The result of *potholes* becoming united in this manner is a complex depression consisting of concave surfaces, generally with vertical axes, and relict portions of dividing walls, which vary from rounded and subdued to sharp and projecting vertical ridges. However, there should not be a hierarchical relationship among the component, coalesced former individuals; they should be of comparable size and not perched one within another. *Coalesced potholes* are generally large and usually complex, with secondary internal sculpting (Nemec et al., 1982; Jennings, 1983; Lorenc et al., 1994). *Coalesced potholes* can exist in one of three associations:

- *Potholes* within a channel margin coalesce, forming an embayment (Fig. 22A).
- *Potholes* coalesce at a *knickpoint*, a process leading to the upstream migration of the *knickpoint* and the formation of a convoluted *inner channel* (Elston, 1917, 1918; Jennings, 1983; Wohl, 1993, 1999; Wohl et al., 1999; Kunert and Coniglio, 2002) (Fig. 22B).
- *Potholes* occurring on a *bedrock bench* coalesce, form ing a complex closed hollow (e.g., Jennings, 1983:

Figure 21. Large *isolated breached pothole*. Note the sharp vertical rim on the right-hand side and the rounded one on the left (downstream side) and the presence of *spiral furrows*. The walls of this *pothole* still enclose up to 270° of the original feature, hence we can be confident about naming it as a *breached pothole* rather than a *lateral pothole*. This origin is also suggested by the shaft-like morphology, the *spiral furrows*, and the pattern of sharp and rounded rims. This example is now active only infrequently, as shown by the tree growing inside it. Nam Mae Chaem (Ob Noi), Thailand. Upstream to right. Granitic gneiss.

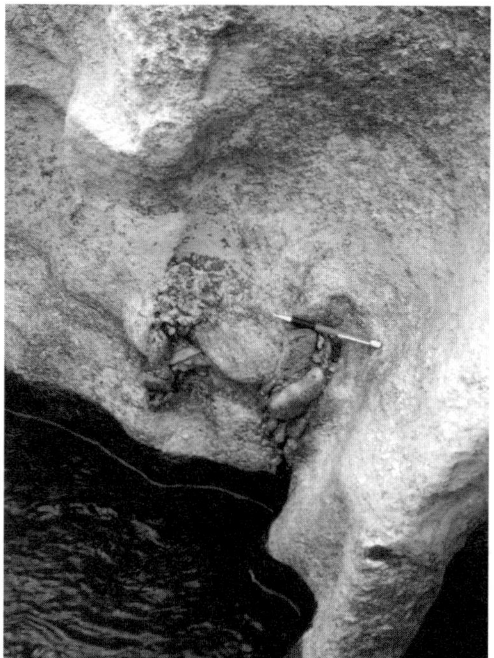

Figure 20. *Coalesced potholes* with relict sediment fills. Nam Mae Chaem (Ob Luang), Thailand. Granitic gneiss. Pen for scale.

Figure 22. A: *Coalesced potholes* forming an embayment within a channel margin. Nam Mae Chaem (Ob Luang), Thailand. Upstream to left. Granitic gneiss. The field of view is ~3 m wide. B: *Inner channel* forming through knickpoint retreat resulting from the coalescence of *potholes*. River Dee, UK. Upstream top right. Limestone. The width of the *inner channel* in the bottom left-hand corner of the photograph is ~4 m

Plate 6 therein; Kale and Shingade, 1987; Lorenc et al., 1994: Fig. 3D therein). A group of *potholes* in the process of coalescing can become interconnected while leaving large portions of their walls and rims intact, producing an outcrop with a "Swiss cheese" appearance. *Potholes* at an advanced stage of coalescence often produce a surface of rounded depressions representing the floors of former *potholes*, separated by a network of ridges and spires that are the remnants of the dividing walls of the former *potholes* (Fig. 22A). The spires and ridges tend to become smooth and convex on the upstream side (i.e., they develop into *upstream-facing convex surfaces*; section 2.2c) and may become undercut on the downstream side, giving them a hooked profile.

(iii) *Natural arches* and *natural pillars*: These are found in *breached* and *coalesced potholes* in which breaching of the wall first took place below the rim of the *pothole* and never progressed to removing a whole side of the *pothole* wall. A small part of the wall on the breached side was left intact, under which water could flow out of the *pothole*. They are termed either *arches* or *pillars* depending on the orientation of the remnant limb (i.e., whether horizontal or vertical) (Figs. 23 and 24). *Natural arches* are described by Blank (1958), Nemec et al. (1982), Jennings (1983), and Kale and Shingade (1987), and illustrated by Lorenc et al. (1994: Figs. 5C and 5D therein). The formation of larger-scale (on the scale of the channel width) *natural arches* is considered in section 6.2i.

(e) *Lateral potholes*. This name was first applied to incomplete potholes in near-vertical faces by Zen and Prestegaard (1994). They are rounded depressions in channel sidewalls and in the flanks of obstacles, and show evidence of both horizontal and vertical erosion. They range from nearly circular (as viewed horizontally, facing the sidewall), concave surfaces to elongate, vertically orientated *furrows*. They may or may not be overhanging in the upper part, and the floor is also variable in character (see below). Zen and Prestegaard (1994) described how groups of *lateral potholes* occur preferentially within the region of vortical flow in shear zones along the flanks of obstacles to the flow such as rock islands. Groups of *lateral potholes* also develop on the lee sides of large obstacles. On a smaller scale, a single *lateral pothole* is often formed within the flow separation

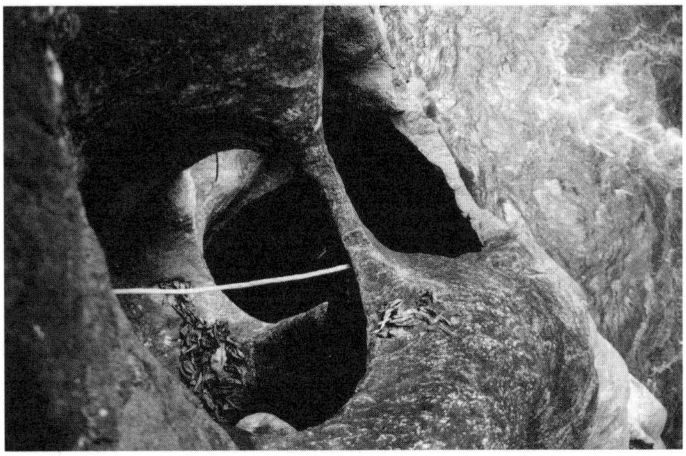

Figure 23. *Natural arch*. Nam Mae Chaem (Ob Luang), Thailand. Upstream bottom right. Granitic gneiss. The *pothole* to the left of the arch is ~1 m across.

Figure 24. *Natural pillar*. Nam Mae Chaem (Ob Luang), Thailand, looking upstream. Granitic gneiss. The *pillar* is ~1.5 m high.

cell in the lee of a projection or obstacle; such *lateral potholes* are equivalent to the "leeside corner potholes" of Ängeby (1951).

Zen and Prestegaard (1994) noted that *lateral potholes* are not simply *breached potholes*, as indicated by three lines of evidence:

(i) Their rims are rounded and integral parts of the concave surface of the *lateral pothole*, indicating development in conjunction with the surrounding rock surface.

(ii) If they have been breached, large amounts of rock must have been removed (from the surrounding surface) without reshaping the interior of the *pothole*, since the arc of the interior is often a small fraction of a full circle. Furthermore, this would make the original complete *potholes* very large, and such large *potholes* would have been interfering, i.e., there is insufficient space available for them to have been originally complete.

(iii) The occurrence of *hierarchical compound lateral potholes* (i.e., secondary ones inside primary ones; see below) essentially precludes the possibility that they were initially discrete forms.

However, the first of these statements is not true in all cases. The presence or absence of sharp rims is not conclusive evidence for the walls of a bedform having been breached. Sharp edges can arise within, around, and between a variety of bedforms that have never been breached (see Discussion). Some bedforms with sharp rims have been observed by the authors that are otherwise typical of *lateral potholes*, and with an internal morphology dissimilar to complete *potholes*, and therefore dissimilar also to *breached potholes*; these sharp-rimmed features are therefore included with *lateral potholes*. Furthermore, *lateral potholes* can occur in conjugate form (Fig. 29), as secondary features inside large primary *potholes*, and contiguously with other *concave features*, and sharp vertical crests may be present in all these circumstances. It is also likely that with time, *breached potholes* have at least one, and perhaps both, of their vertical rims rounded off (see Discussion). Nevertheless, the argument of Zen and Prestegaard (1994), that *lateral potholes* are morphologically and genetically distinct from *breached potholes*, is supported by field observations made for this study. Further to the second line of evidence cited above, it may be noted that to remove a large volume of material from the adjacent rock surface but not from within the *pothole*, which would be necessary if *lateral potholes* were in fact *breached potholes*, is contrary to the normal behavior of scoured *concave features*, within which erosion is enhanced relative to the surrounding surface.

As noted above, the internal morphology of *lateral potholes* generally differs from both complete and *breached potholes*. *Lateral potholes* are very variable, and it is easier to say what they are not than what they are. Their walls do not generally describe arcs of circles in horizontal sections (Zen and Prestegaard, 1994). The relief of *lateral potholes* is typically less than their width (i.e., their streamwise dimension), and their walls rarely enclose more than 180° of a complete turn, so that they normally narrow monotonically with depth into the rock face from the aperture. This contrasts with *breached potholes*, which tend to form deep embayments enclosing more than 180° of a circle. As a result of both their low relief and lack of circularity, *lateral potholes* are not cylindrical and shaft-like and do not have the appearance of vertically sectioned complete *potholes*. This is especially the case where they are overhanging and/or *equant* (see below). Many *lateral potholes* also taper upwards, forming a teardrop shape (e.g., Gjessing, 1967; Springer and Wohl, 2002).

Lateral potholes may exist with *furrows* of various types (*parallel-sided furrows*, *transverse furrows*, *oblique sloping furrows*, *current crescents*, *furrow complexes*; see sections 1.2c, 1.3a, 1.3b, 1.4, 3.1a) leading smoothly into them from above, and connected to them (Fig. 25), illustrating that they are closely related to these various types of *furrows*. Complete *potholes* do not show this relationship, which differs from that between *potholes* and their *entry* and *exit furrows* in that the latter are subordinate, passive markers of the flow induced by the *pothole* and are incidental to the *pothole*, except where the *furrows* become very large. However, the *furrows* associated with *lateral potholes*, where present, are of dimensions equal to or greater than the *lateral pothole* and are an integral, active part of the system. In fact, *lateral potholes* form a continuum with these types of *furrows* in terms of form and of orientation.

Wohl and Achyuthan (2002) describe *lateral potholes* from an unnamed river in India, and examples may be seen in the illustrations of Thybony's (2000) description of *slot canyons*. *Lateral potholes* have been described from a cave setting by Springer and Wohl (2002), who also inferred the existence of recirculating, downward-spiraling vortical flow from an analysis

Figure 25. *Longitudinal furrows* (top half of photo) leading into and merging with *lateral potholes* (lower half of photo). Looking upstream. Nam Mae Chaem (Ob Luang), Thailand. Granitic gneiss. The notebook is 15 cm long.

Figure 26. A *joint furrow* can be seen running from top to bottom, with a *pothole* (B) in the process of being breached by it. Just below center on the opposite side of the *joint furrow* is an *incomplete pothole* (I). It is impossible to tell whether it was breached at an earlier time, or whether it is a *lateral pothole*. Two *coalesced potholes* (C) can be seen left of center. River Dee, UK. Flow from top to bottom. Limestone. Penknife for scale.

of *scallop* patterns within the bedforms. *Lateral potholes* have also been described previously under different names. Maxson and Campbell (1935) described elongated varieties of *lateral potholes* but considered them to be a type of *flute*. Such a classification is not supported in this report. In many descriptions, a clear distinction between *potholes* that have always been incomplete (i.e., *lateral potholes*) and those that were originally complete, but that were subsequently breached, has been lacking. All features that lack one sidewall have often been grouped together as "half potholes," without consideration of their origin (e.g., Ängeby, 1951; Whipple et al., 2000a). Wohl and Ikeda (1997) observed *lateral potholes* evolving at the meanders in a sinuous experimental channel and described them as "breached potholes," although they had never been complete (see section 1.2f). The "side holes" of Sato et al. (1987) and the "horizontal holes" of Lorenc et al. (1994) are probably varieties of *lateral potholes*. *Lateral potholes* are known to occur in subglacial environments as a type of s-form (Gjessing, 1967; Kor and Cowell, 1998), and some of the semicircular features developed experimentally by Shepherd and Schumm (1974: Fig. 4 therein) are similar to *lateral potholes*. *Incipient lateral potholes* also occur (see section 1.1a part vi, above).

While many examples of *breached* and *lateral potholes* can be clearly distinguished, others cannot (e.g., Fig. 26), and each example must be assessed individually on its own merits. This is an example of the convergence of form, discussed in a later section. In the absence of clearer guidelines and better understanding of *pothole* evolution and flow structures, the identification of these bedforms involves some subjectivity. It is suggested here that the neutral term *incomplete pothole* should be applied to indeterminate cases.

Zen and Prestegaard (1994) identified two types of *lateral potholes* based on the nature of the *pothole* floor. They also described compound types, and to these are added further definitions by the present authors. There is a continuum in form between enclosed *potholes* and *lateral potholes* as the slope of the channel margin increases. In all types listed below, the back wall may or may not be overhanging in the upper part.

(i) *Closed lateral pothole*: The floor contains a closed basin, such that the floor within the *pothole* is lower than the lowest point of the rim on the open side (Fig. 27). Such features were named "conical lateral potholes" by Zen and Prestegaard (1994).

(ii) *Open lateral pothole*: The floor does not contain a closed depression, and the floor is everywhere sloping outwards toward the open side (Fig. 28A). Such features were named "bucket-seat lateral potholes" by Zen and Prestegaard (1994). It seems logical to suppose that *open lateral potholes*, which in many cases are rather shallow forms, generally represent juvenile features, whereas *closed lateral potholes*, which require a greater relief and a greater volume of rock removal for the same rim circumference, represent mature features, and that with time the former will often evolve into the latter. However, this will also depend on the slope of the surface in which the *lateral pothole* is

Figure 27. *Closed lateral potholes*. Each *pothole* has a closed basin in its floor. These *lateral potholes* are also *equant*, and most are *sharp-rimmed*. Nam Mae Chaem (Ob Luang), Thailand. Upstream to left. Granitic gneiss. The field of view is ~6 m wide.

developing; *closed* forms will arise more quickly on gentler slopes.

(iii) *Equant lateral potholes* and *linear lateral potholes*: All of the examples illustrated by Zen and Prestegaard (1994) had widths and heights of comparable magnitude. These may be described as *equant lateral potholes* (Fig. 27). However, many elongated forms also occur that are much greater in height than in width. These *linear lateral potholes* are essentially vertically orientated *transverse furrows* (Fig. 28) and were described by Maxson and Campbell (1935) from examples in the Grand Canyon. Each *linear lateral pothole* tends to be rather uniform in width, and to lack a well-defined floor, while *equant lateral potholes* do have a well-defined floor if they are of sufficient relief. *Linear lateral potholes* are common both as secondary forms within large *potholes* and as conjugate forms in near-vertical channel margins (Fig. 29). The highly elongated form of many *linear lateral potholes* argues against an origin through the breaching of complete *potholes*, since this would normally occur through an increase in diameter rather than depth. There is of course a continuum in form between *equant* and *linear lateral potholes*, and *lateral potholes* of any proportions may taper upwards.

(iv) *Sharp-rimmed lateral potholes*: similar to other *lateral potholes*, except for possessing sharp rims, even when isolate (Figs. 27 and 28A). Either or both of the vertical rims may be sharp. Alexander (1932: Fig. 11 therein)

Figure 28. A: *Lateral pothole*. This example has an *open*-type base; the feature has no basin in its floor, and the internal surface everywhere slopes outwards. The *lateral pothole* is at least twice as high as it is wide, and it is therefore described as *linear*. It is also *sharp-rimmed* on both sides. Nam Mae Chaem (Ob Luang), Thailand, looking upstream. Granitic gneiss. The notebook is 15 cm long. B: *Lateral potholes*. These examples are highly *linear*. Both *sharp-rimmed* and rounded varieties are present. Nam Mae Chaem (Ob Luang), Thailand. Upstream to left. Granitic gneiss. The vertical wall containing the bedforms is ~2.5 m high at the right-hand side.

illustrates a *lateral pothole* with one sharp and one rounded vertical rim.

(v) *Compound lateral potholes*: Zen and Prestegaard (1994) described two varieties of *compound lateral potholes*, those with a nested arrangement of smaller secondary *lateral potholes* within a larger primary one, often in perched positions on the sidewall (referred to herein as *hierarchical lateral potholes*) (Fig. 30), and those consisting of a series of linked small "basins." These secondary basins would probably come under the category of *short furrows*. A further type of *compound lateral pothole* that may be found is that of *paired lateral potholes*, consisting of two equal-sized and closely related *lateral potholes* (Fig. 31). The morphology, especially the distribution of sharp and rounded edges, of both *hierarchical* and *paired lateral potholes* suggests that the component *potholes* contain concordant, counterrotating vortices when active (see Discussion). The nested arrangement of the component forms allows *hierarchical lateral potholes* to be easily distinguished from *coalesced potholes* and conjugate *lateral potholes*, in which neighboring hollows are of similar magnitude and do not have a nested or perched relationship. It should be noted also that many of the secondary holes in complete *hierarchical potholes* are *lateral potholes*.

(vi) *Spiral-furrowed lateral potholes*: Some of the *linear lateral potholes* described by Maxson and Campbell (1935) had *spiral furrows* and a *central boss* in the floor. These were referred to mistakenly as "spiral flutes."

Figure 29. Conjugate *linear lateral potholes*. Note the vertically orientated, ridge-like projections, which are the crests between contiguous individuals. Many of these crests are sharp. Nam Mae Chaem (Ob Luang), Thailand. Upstream to right. Granitic gneiss. Field of view is ~5 m wide.

1.2 Longitudinal Furrows

These are *furrows* orientated parallel to the local flow direction. Although in general this direction is similar to the orientation of the channel as a whole, the *furrows* and the local flow may deviate markedly from the general direction of the channel, and thus, provided they can be recognized, *longitudinal furrows* observed at low stage on exposed surfaces can be used to reconstruct the often complex flow patterns that exist within bedrock channels at formative discharges. However, in cases where the local flow deviates greatly from the general direction of the channel, and consists of upstream-directed currents

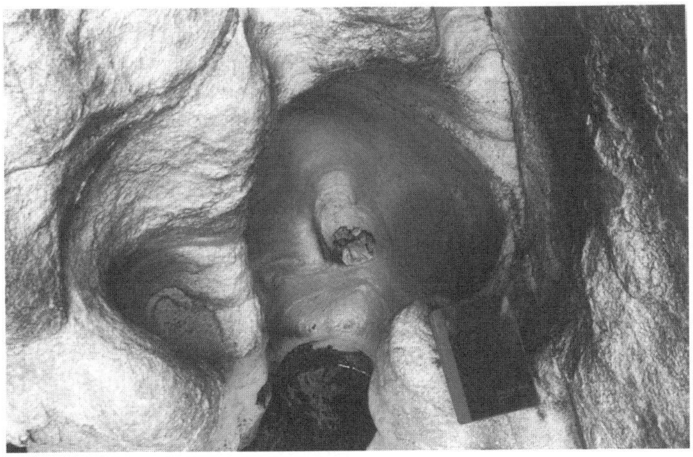

Figure 30. *Compound lateral pothole* of the *hierarchical* variety. Note the nested arrangement of small secondary *potholes* within a larger primary one. Nam Mae Chaem (Ob Luang), Thailand. Upstream to left. Granitic gneiss. The notebook is 15 cm long.

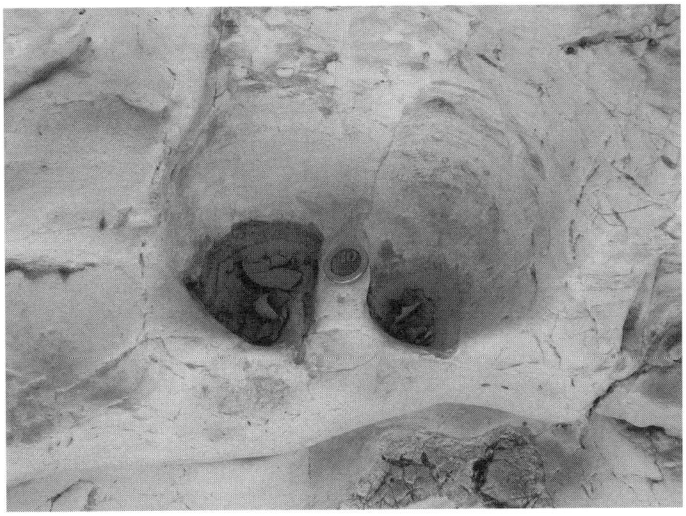

Figure 31. *Paired lateral potholes*. Wadi Hazazon, Israel. Upstream to right. Dolomite. Coin for scale.

within attached flow separation eddies in the lee of boundary irregularities, the term "longitudinal" is no longer applied. *Furrows* eroded by such currents are assigned to the group of *reversed furrows* (section 1.3c). Thus, *longitudinal furrows* satisfy two criteria: They are parallel to the local flow direction, and this direction does not deviate greatly from the general orientation of the channel.

Longitudinal furrows comprise a spectrum of *furrow* types exhibiting a wide range of morphologies and length:width ratios (Fig. 32). It should be noted that the bedforms described below are not discrete types, and there is continuity of form, not only between the members of this group, but also between this group and *potholes*, as the length:width ratio is reduced toward unity. Also, it is possible for individual *furrows* simultaneously to possess characteristics of more than one of the types described below, becoming a hybrid form.

(a) *Flutes*. Allen (1971a) discussed *flute* morphology and flow structures in detail and introduced many terms in his descriptions of them. In mud deposits, where they are best known, *flutes* have rims that flare in the downstream direction, eventually becoming indistinguishable from the general surface, producing a more or less parabolic outline in plan view. It is common for *flutes* to contain a low, rounded *median ridge* along their axis, and narrow lateral furrows along the flanks. They have steeply sloping lee faces and more gentle upstream-facing slopes and are deepest near the upstream (proximal) end of the depression.

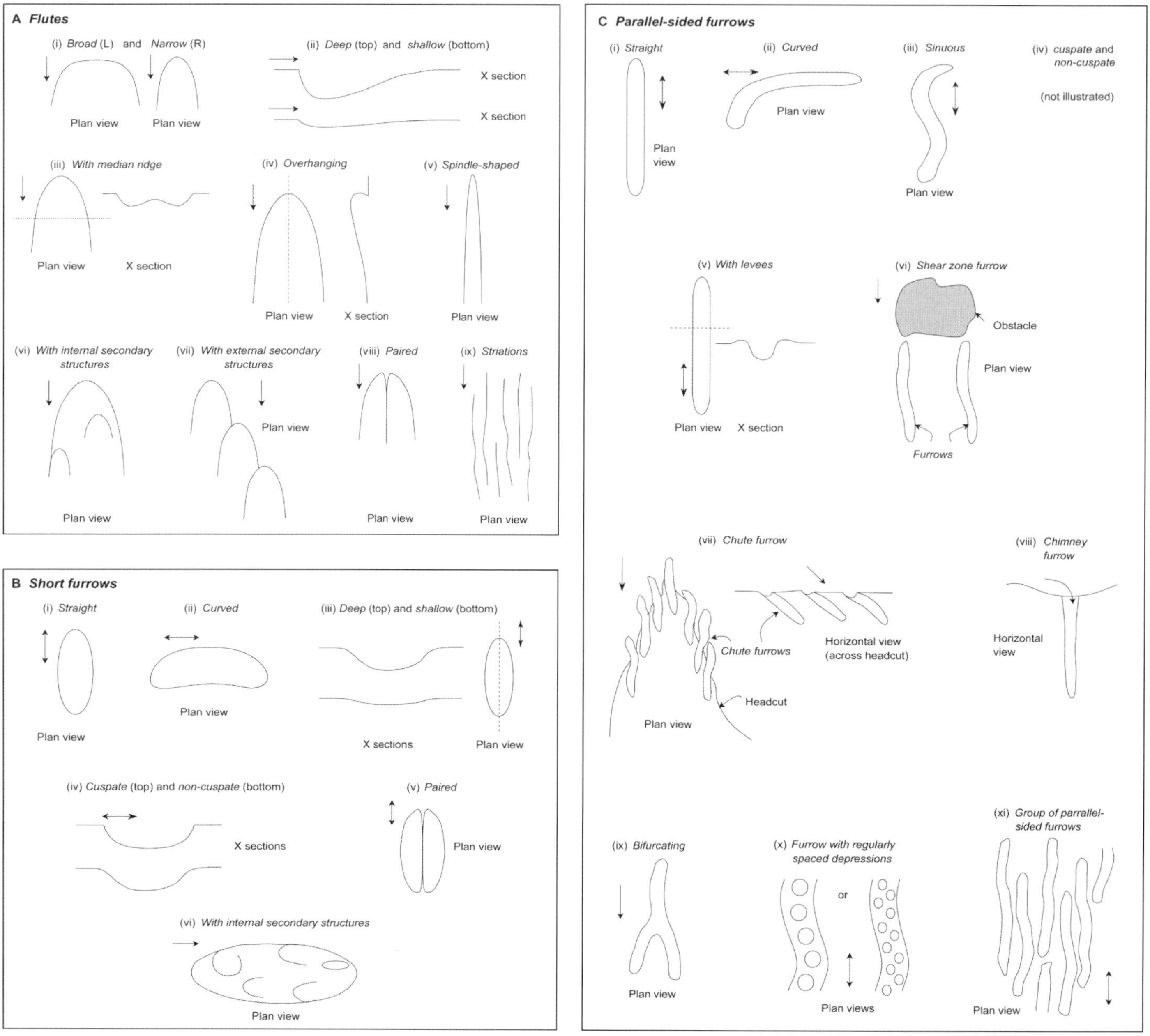

Their rims are cuspate, at least in the proximal region. In the more resistant and heterogeneous substrate of rock, however, *flutes* are often rather different and may only approximate the "ideal" shape described above. They lack lateral furrows, and *median ridges* are rare. They are often of greater depth (relative to width), and their rims may overhang at the upstream end. They are also more irregular in shape than their mud counterparts. The more complex *flute* forms described by Allen (1971a) from mud and plaster of Paris, such as "corkscrew" and "comet-shaped" *flutes*, are absent in bedrock. Furthermore, at some sites there is a lack of truly open *flutes*—those whose rims flare in the downstream direction along the whole length of the bedform. Rather, their rims, though becoming less distinct with distance from the proximal end, often curve inwards again to join up at the distal end, i.e., the *flutes* are closed. Such *flutes* are therefore gradational with *short furrows* and further illustrate the idea of continuity of form. In simple terms, these differences between *flutes* in mud and rock are probably related to the fact that the compliant medium of mud allows for the fullest expression of the flow hydrodynamics to be recorded, while the more resistant and heterogeneous medium of rock does not. It appears that open *flutes* are favored by high-velocity flows (e.g., examples from the Colorado River [Maxson and Campbell, 1935], Ganges [Hancock et al., 1998], and Burdekin Gorge [Wohl, 1992b]). Both open and closed *flutes* are also known to occur in previously glaciated regions as s-forms (Kor et al., 1991). The term *flute* is often used in a broad sense to refer to any kind of *longitudinal furrow*. However, the authors favor a stricter definition of *flutes* following the description just given, and as discussed in the terminology section. *Flutes* (sometimes referred to as "troughs" or "spindles") are recorded in a fluvial bedrock environment by Maxson and Campbell (1935), Maxson (1940), Allen (1971a), Baker and Pickup (1987), Wohl (1992b), Tinkler (1993, 1997a), Baker and Kale (1998), Hancock et al. (1998), Wohl and Ikeda (1998), Gupta et al. (1999), and Whipple et al. (2000a). It should be noted, however, that Maxson and Campbell (1935) described several different bedform types under the heading of *flutes*, of which only their "pocket flutes" may be justifiably referred to this group.

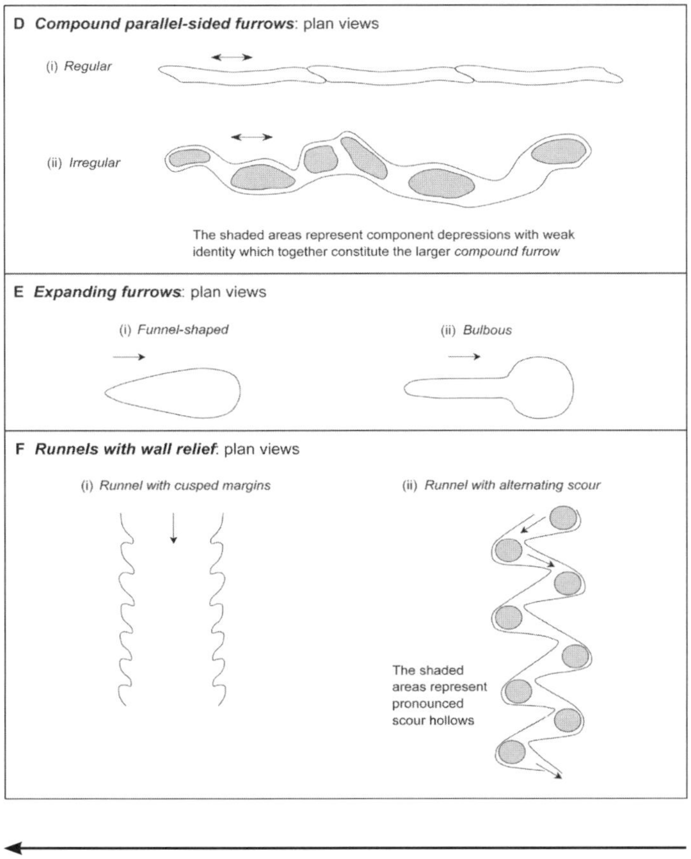

Figure 32. *Longitudinal furrows*. Arrows indicate flow directions (where appropriate). Dashed lines indicate lines of cross sections.

Flutes require some form of "defect" that generates flow separation for their initiation (Allen, 1971a). In cohesive sediment, this may be a burrow or some piece of extraneous material. However, in bedrock this defect is usually a topographical irregularity such as a corner or projection, a sharp (though not necessarily large) convex change in orientation of the bedrock surface, or some heterogeneity within the rock that either is resistant and stands proud or is weak and quickly eroded away. The defect required to initiate *flutes* may be a preexisting bedform; the *flute* is then secondary sculpting within a primary feature. The initiation of *flutes* at relatively minor changes in surface orientation (say a few degrees) indicates that they do not always originate at obvious defects. *Flutes* occur preferentially in regions of flow expansion, such as toward the rear of boulders (Hancock et al., 1998; Whipple et al., 2000a). These are regions of adverse pressure gradient, a condition that promotes flow separation (Tritton, 1988: p. 127, 141–145) and therefore *flute* development.

Hancock et al. (1998) argue from field observations that *flutes* migrate upstream, whereas in fact the leading edge of a *flute*, whether in bedrock, mud, or plaster of Paris, may migrate upstream or downstream, depending on the angle of the proximal slope and the relative rates of erosion on this slope and in the area outside the *flute*. For instance, in Allen's (1971a) theoretical treatment of the development of the longitudinal profile of a *flute*, a profile was generated in which the leading edge migrated upstream, whereas in his experiments, examples of *flutes* migrating both upstream and downstream were generated. Geometrical considerations indicate that the steeper the proximal slope, the more likely it is that the leading edge of the *flute* will migrate upstream.

Flutes may occur as isolate or conjugate forms. Allen (1971a, 1971b) showed conclusively that the origin of the *solutional forms* in caves known as *scallops* (section 4.2) lies in the growth and interference of initially isolate *flutes*, during which process the *flutes* become fully conjugate. However, it was argued in the terminology section that the conjugation of *flutes* is a process rather than an event, since *flutes* may be partially conjugate while still essentially retaining the character of the original isolate, more or less parabolic scour hollows. The conjugate *flutes* illustrated by Maxson and Campbell (1935) and Maxson (1940: Fig. 1 therein) are quite distinct from *scallops*, for instance. *Scallops* are ripple-like forms, which may be highly three-dimensional or more two-dimensional in character, and whose origin in initially isolate *flutes* is normally not obvious. The differentiation of conjugate *flutes* and *scallops*, as described in the terminology section, is adopted here.

The *flutes* described below are all approximately of the parabolic form of Allen (1971a), with the exception of *spindle-shaped flutes*.

(i) *Broad* and *narrow flutes*: These descriptors are close in meaning to those of Allen (1971a). A *broad flute* is as wide or wider than it is long, while a *narrow flute* is narrower than it is long (Figs. 33 and 34). A *broad flute* has a rim that, in the proximal region, is only slightly concave in the downstream direction (in plan view), while a *narrow flute* has a highly concave rim.

Figure 33. *Broad flute*. This example is slightly skewed with respect to the flow direction, perhaps due to some structural influence. Wadi Hazazon, Israel. Flow from top left to bottom right. Limestone. Coin for scale.

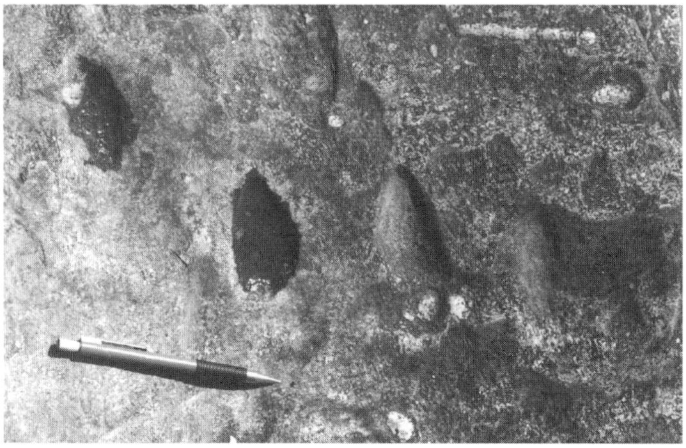

Figure 34. *Narrow flutes*. Nam Mae Khlang, Thailand. Flow from top to bottom. Granitic gneiss. Pen for scale.

(ii) *Deep* and *shallow flutes*: *Flutes* show a great range in depth from a very small fraction of their length to a depth approaching their length. Such a wide range in profile shape must give rise to markedly different flow structures in these two extremes. On this basis, they are arbitrarily divided into *deep* (depth more than a quarter of their length) and *shallow flutes*. The *flutes* shown in Figures 33 and 34 are *shallow*, while that shown in Figure 37 is *deep*. *Shallow flutes* are analogous to the subglacial s-forms known as muschelbrüche (Kor et al., 1991; Shaw, 1996; Kor and Cowell, 1998). Bryant (2001: p. 86) described *shallow flutes* (which he also named muschelbrüche) from coastal environments due to tsunamis.

(iii) *Flute with median ridge*: *Median ridges* (Fig. 35) were found by the authors only in a few *broad flutes*, although they are identified by Allen (1971a) as a very common element of parabolic *flute* morphology in mud and plaster of Paris. *Flutes with median ridges* are analogous to the s-forms known as sichelwannen (Kor et al., 1991; Shaw, 1996; Kor and Cowell, 1998).

(iv) *Overhanging flute*: Maxson and Campbell (1935) and Hancock et al. (1998) describe *flutes* whose rims overhang at their proximal ends. All these authors consider that such *flutes* develop from bedforms that initially do not overhang, but in which the upstream rim becomes progressively steeper as a result of intense erosion. From geo-

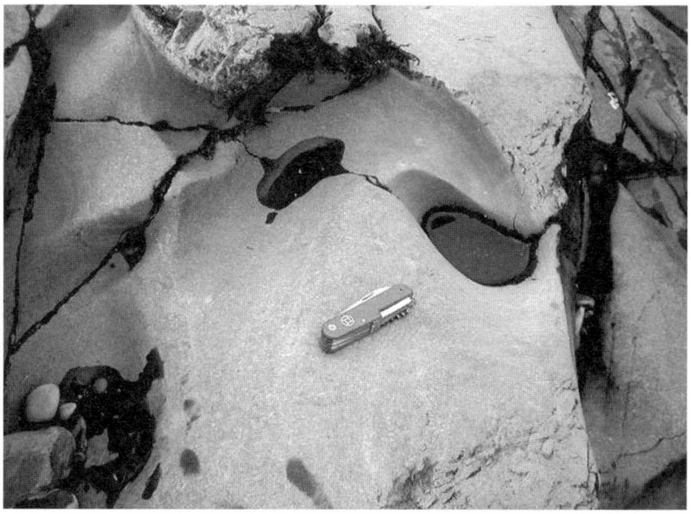

Figure 35. *Flute with median ridge* and *internal secondary structures*. The penknife is positioned on top of the *median ridge*, and two *short furrows* have been eroded into the principal furrow of the *flute*. River Lune (Tebay), UK. Flow from top to bottom. Calcareous mudstone.

metrical considerations, it can be shown that *overhanging flutes* must migrate upstream as they erode downwards. Hancock et al. (1998) use these features to advance their argument that erosion is most intense on the lee sides of bedforms and obstacles that generate separated flows. However, *flutes* that overhang appear to be the exception rather than the rule, and Allen (1971a) clearly demonstrated that the highest rates of erosion within *flutes* occur at and just downstream of the point of reattachment, on the stoss rather than the lee slope of the *flute*.

(v) *Spindle-shaped flute*: Allen (1971a) described *flutes* that are much longer than wide, with sharply pointed proximal ends and without a median ridge, as *spindle-shaped*. These are one of the smallest bedforms that occur in bedrock, ranging in size from ~10 cm down to a few millimeters in length (Fig. 36). They are scour marks that extend downstream from small "defects" in the rock (such as vesicles, fossils, or large individual grains). These defects are probably smaller than the critical size that defects must exceed if the *flutes* to which they give rise are to grow indefinitely. Defects below this critical size are known to give rise to *spindle-shaped flutes* that undergo an initial period of growth, but both defect and *flute* are subsequently erased by general bed lowering in plaster of Paris (Allen, 1971a; see also *lineations*, below). Baker and Pickup (1987) described *spindle-shaped flutes* in the Katherine Gorge of Australia, where they were found to originate at constituent mineral grains. They are also sometimes referred to as "pinheaded rattails" (Tinkler and Wohl, 1998a) and appear to be equivalent to the "fret flutes" of Maxson and Campbell (1935). However, *spindle-shaped flutes* are also known to occur on a large scale as s-forms (Kor et al., 1991; Shaw, 1996).

(vi) *Flute with internal secondary structure(s)*: Allen (1971a) described how mature, isolate *flutes* may give rise to secondary *flutes* within their boundaries, either close to their rims proximally or close to their central axes in the distal region. *Secondary structures* of the first type are illustrated in Figure 37. However, *secondary structures* in *flutes* in bedrock are apparently more diverse than in mud and plaster of Paris. For example, *potholes*, *short furrows*, and *flutes* may be found as *secondary structures* in any part of mature (i.e., large) primary *flutes* (e.g., Fig. 35), although they are more common adjacent to the rim. The growth of *internal secondary flutes* both at the rim and in the axial region of a primary *flute* may be seen in Hancock et al. (1998: Fig. 11 therein) from the River Indus, Pakistan. A further type of *secondary structure* may be found in mature *deep flutes*. Here this is named *irregular sculpting* (section 2.2d), and it is a general phenomenon of the lee faces of many rock projections and nontransported boulders. Mature *deep flutes* effectively present a large lee face at their upstream (proximal) end and therefore become subject to this type of sculpting.

(vii) *Flute with external secondary structure(s)*: In addition to those found within primary *flutes*, *secondary structures* may also be generated outside but adjacent to a primary *flute*. This may occur at a linear "defect" in the rock surface, such as a sharp ridge, as illustrated in Figure 38. Hancock et al. (1998: Figs. 3 and 4 therein) illustrate a similar phenomenon occurring at a vein. The defect initiates the first *flute*, the growth of which modifies the outline of the

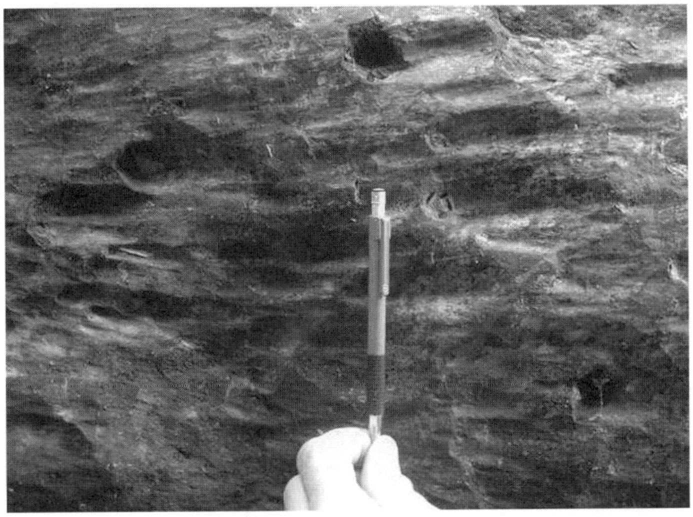

Figure 36. *Spindle-shaped flutes* developing at the sites of vesicles in volcanic rock. Huai Nang Rong, Thailand. Flow from left to right. Rhyolitic agglomerate.

Figure 37. *Flute with internal secondary structure*. A new *flute* (arrowed) has been developed within a preexisting one close to its rim at the proximal end. Overall, the appearance of the outline of the two combined structures retains that of a single *flute*. Borrow Beck, UK. Flow from right to left. Calcareous mudstone. Pen for scale.

defect. It appears that an *external secondary flute* is subsequently initiated at the corner formed by the intersection of the rim of the primary *flute* and the line of the defect. Growth of the *secondary structure* then modifies the outline of both the defect and the primary *flute*. However, it appears that *external secondary flutes* may also be generated in the absence of such a linear defect. In this case, the *secondary flutes* are generated close to the primary *flute(s)*, in a characteristic position relative to them, this being downstream and to one side. Thus, assemblages of closely spaced and conjugate *flutes* are created with a characteristic diagonal alignment of individuals. These may be termed en echelon assemblages and can be seen in the illustration of Maxson (1940: Fig. 1 therein) and in Figure 39 of this report. Kor et al. (1991) and Shaw (1996) report en echelon assemblages of sichelwannen. This alignment of individuals is similar to, though less regular than, that which exists in rhomboid ripples (Hoyt and Henry, 1963; Otvos, 1965), although *flute* crests are convex-upstream in plan view, whereas rhomboid ripple crests are convex-downstream.

(viii) *Paired flutes*: *Narrow flutes* can sometimes be found in a paired state. The outline of the pair seen as a whole has the appearance of a single *flute* but clearly contains within it two component *flutes* that are slightly divergent in orientation, and that are separated by a central ridge (Fig. 40). If the pair is viewed as a single bedform, the central ridge can be seen as analogous to a *median ridge*, except that it is joined to the proximal rim of the bedform and therefore completely divides the component *flutes*.

(ix) *Lineations*: These were found at only two sites by the authors: the River Dee and Lan Rak. They are small-scale *ridges* and *furrows* orientated parallel to the flow with a transverse spacing of between a few millimeters and ~2 cm. Their length is indeterminate but is on the order of a few centimeters (Fig. 41). They possess no definite head or tail but merge with the surrounding rock surface at either end. The *ridges* and *furrows* are rounded in cross section. In most places, the origin of these *lineations* is not

Figure 38. *Flute with external secondary structures*. A row of small *flutes* has developed along a linear defect. It is interpreted that a primary *flute* developed at some point along the defect, which subsequently spawned other *flutes* formed along the defect as secondary features, adjacent to the primary *flute* (see text for explanation). River Dee, UK. Flow from top to bottom. Limestone. The top scale of the ruler is in centimeters.

Figure 39. *En echelon flutes*. Nam Mae Chaem (Ob Luang), Thailand. Flow from bottom right to top left. Granitic gneiss. The rock projection is ~4 m across.

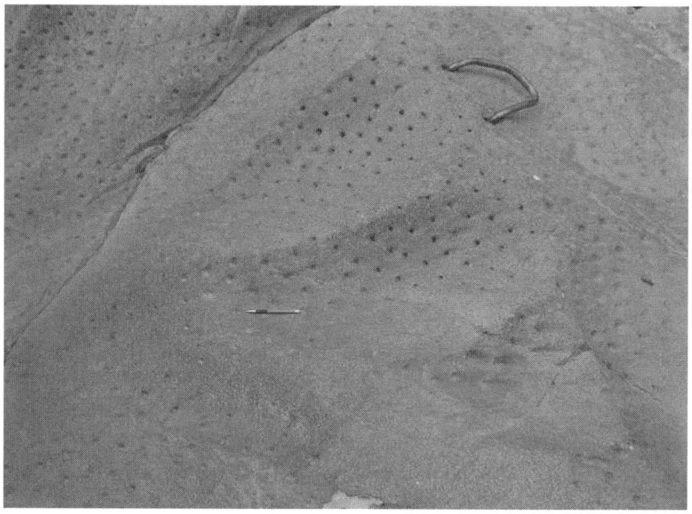

Figure 40. *Paired flutes* (of *narrow* form). The rock has been artificially drilled with many small holes as part of engineering works. Nam Mae Chaem (Ob Luang), Thailand. Flow from bottom left to top right. Granitic gneiss. Pen for scale.

clear. However, sometimes it can be seen that the *ridges* and *furrows* form the distal parts of *spindle-shaped flutes* that occur at small defects in the rock. It appears that these defects are below the critical size for indefinite enlargement (Allen, 1971a), and that with time the defect and its associated *flute* are erased from the bed, leaving only remnant rounded *ridges* that were formerly the distal parts of the *flute* rim. This corresponds to growth pattern II of *flutes* observed by Allen (1971a).

(b) *Short furrows.* A depression with an elliptical rim in plan view and whose length is more than twice its width is here referred to as a *short furrow*. There is no natural break in form between *potholes* and *short furrows*, and this definition is entirely arbitrary. *Short furrows* are not as deep as they are long, and as they do not have the circularity of *potholes*, it is likely that they also do not have the vertically orientated vortices that are capable of drilling deep *potholes*. Instead, they may have a horseshoe vortex, similar to those within *flutes* (Allen, 1971a). They may usefully be compared with *flutes*, to which they are similar, except that in *short furrows* the upstream end is a reflection of the downstream end, whereas the downstream end of a *flute* is indistinct, and the *flute* merges with the surrounding bedrock. Seen in plan view, *short furrows* may have pointed or rounded ends. In cross section, they are smooth and rounded internally, although the rims may be cuspate. These bedforms are small depressions, typically up to ~0.4 m long, and exceptionally up to a meter or so. It should be emphasized that the term *short furrow* does not refer to the absolute length of the bedforms, but rather the length:width ratio, i.e., the elliptical shape of the bedforms in plan view. Jennings (1983) describes features that could be classified as *short furrows* but erroneously refers to them as "gouge holes" (after Alexander, 1932) and ascribes to them flow structures that are apparently without observational or theoretical basis. *Short furrows* are analogous to the s-form known as "closed spindles" (Kor et al., 1991; Shaw, 1996). Karcz (1973) also described elliptical scours in sedimentary environments, although these were on a large scale, with lengths of meters to tens of meters.

(i) *Straight short furrow* (Fig. 42).

(ii) *Curved short furrow*: Although *short furrows* are generally straight, curved examples also occur (Fig. 43) and are common in areas of locally complex topography; they sometimes occur as secondary sculpting in the floors of *potholes*, for example. They may be isolate or contiguous with other bedforms. When they occur as secondary sculpt-

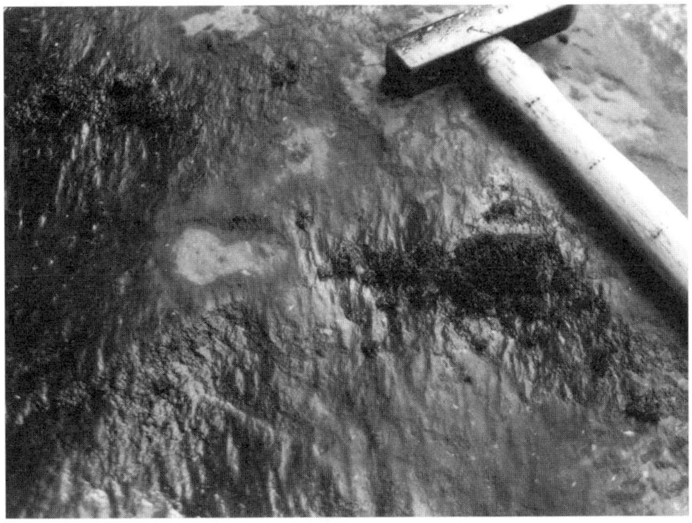

Figure 41. *Lineations*. River Dee, UK. Flow from top to bottom. Limestone. The marks on the hammer shaft are 5 cm apart.

Figure 42. An assemblage of *straight short furrows*. These *short furrows* are of the *shallow* variety. River Dee, UK. Flow from right to left. Limestone. Penknife for scale.

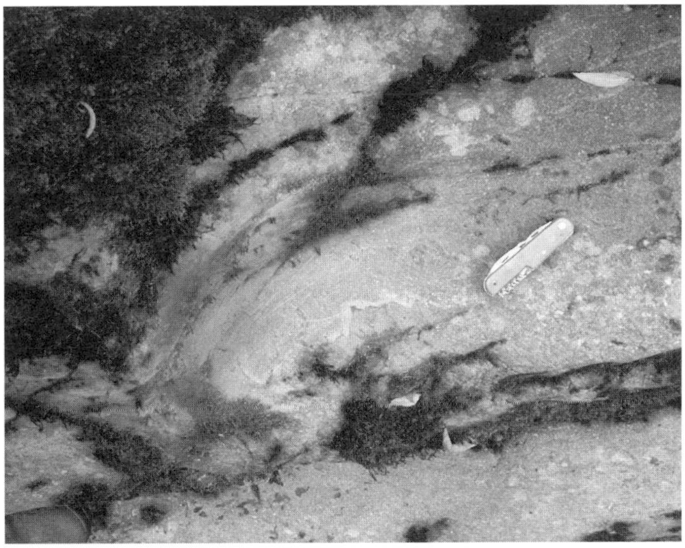

Figure 43. *Curved short furrow*. River Lune (Tebay), UK. Flow from left to right. Calcareous mudstone. Penknife for scale.

ing within a larger bedform, they may also form a style of *furrow complex* known as *yin-yang* (section 1.4b).

(iii) *Deep* and *shallow short furrows*: A similar definition regarding depth may be applied to *short furrows* as that applied above to *flutes*. Figure 42 illustrates *shallow short furrows*, while a *deep short furrow* may be seen in Figure 44.

(iv) *Cuspate* and *noncuspate short furrows*: The rims of *short furrows* may be sharp (Fig. 44), or rounded.

(v) *Paired short furrows*: As with *flutes* and *lateral potholes*, *short furrows* may exist within a paired relationship (Fig. 45). The *furrows* exist side by side, separated by a central ridge that is lower than the external rim of the pair. The component *short furrows* may be parallel or slightly divergent in the downstream direction. There may or may not also be an obvious structural cause for the central ridge dividing the *furrows*; it may, for example, follow a joint or vein.

(vi) *Short furrow with internal secondary structures*: Exceptionally, *short furrows* grow to a large size (50–100 cm long) and develop internal secondary sculpting (*flutes* or small *short furrows*) at the upstream end (Fig. 46).

(c) *Parallel-sided furrows.* As the length:width ratio of *concave features* increases, eventually a state is attained whereby the two sides of the *furrow* seen in plan view are parallel over some distance. The outline is then no longer elliptical as it is in *short furrows*, and the cross section is more or less uniform over a significant portion of the feature. Such a bedform is then a *parallel-sided furrow*. These features are internally rounded in cross section. They may be open at both ends or at one end, or be entirely closed depressions. Where they are closed and have cuspate rims, the upstream ends are often similar to those of *flutes* and *cuspate short furrows*. From the work of Allen (1971a), and by analogy with *spindle-shaped flutes*, it may be expected that *parallel-sided furrows* contain pairs of attached quasi-streamwise, counterrotating vortices, but this aspect of the bedforms has not been examined. Ancient *parallel-sided furrows* in bedrock (i.e., preserved in unconformities) have been described by Shrock (1948: p. 230–236). *Parallel-sided furrows* are also known from sedimentary environments; for instance, Friend (1965) and Allen

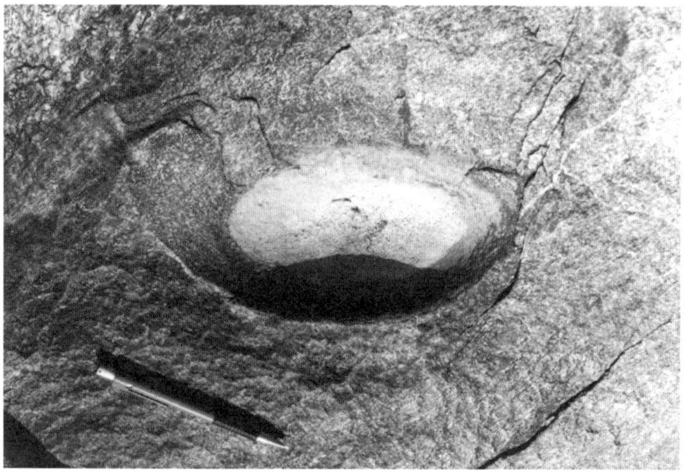

Figure 44. *Cuspate, deep short furrow*. Nam Mae Khlang, Thailand. Flow from left to right. Gneiss. Pen for scale.

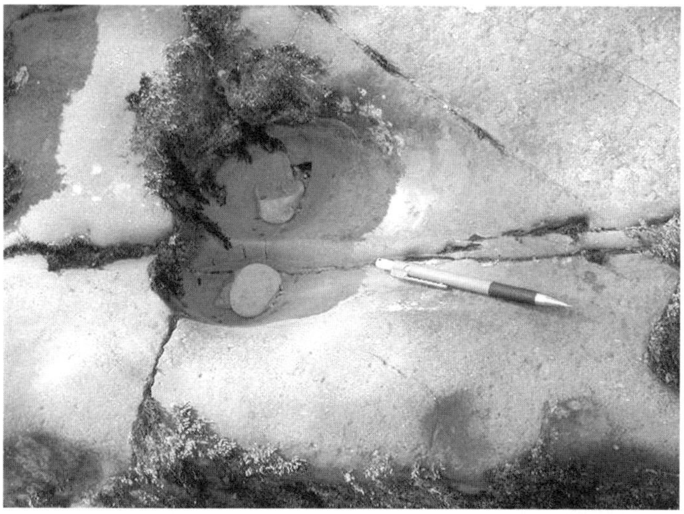

Figure 45. *Paired short furrows* in a parallel arrangement. The central ridge follows a prominent joint. Borrow Beck, UK. Flow from left to right. Calcareous mudstone. Pen for scale.

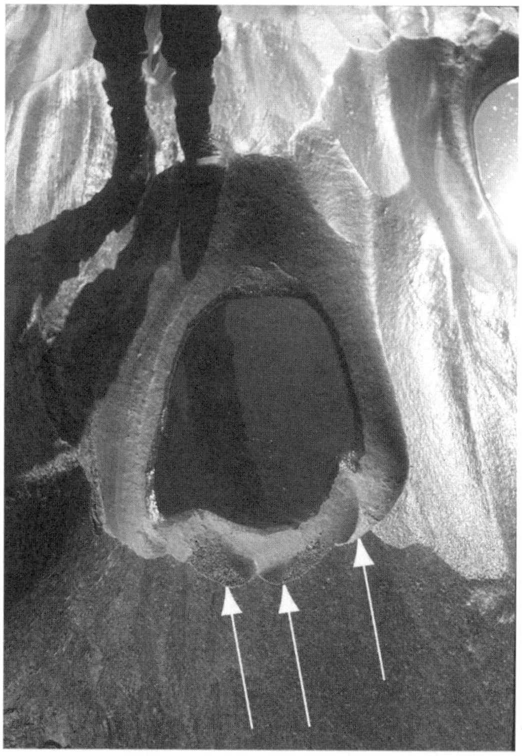

Figure 46. *Short furrow with internal secondary structures*. The secondary features (arrowed) are curved *flutes* at the proximal end of the *furrow*. Watson River, Greenland. Flow from bottom to top. Gneiss. Person (from knees down only) at top of picture for scale. Photo courtesy of Andy Russell.

(1974) describe such features (termed "welts" or "channels") formed in mud in a fluvial environment and preserved in the geological record.

(i) *Straight parallel-sided furrow*: These features are common and are well developed at the lips of *downstream-facing steps* (Fig. 47).

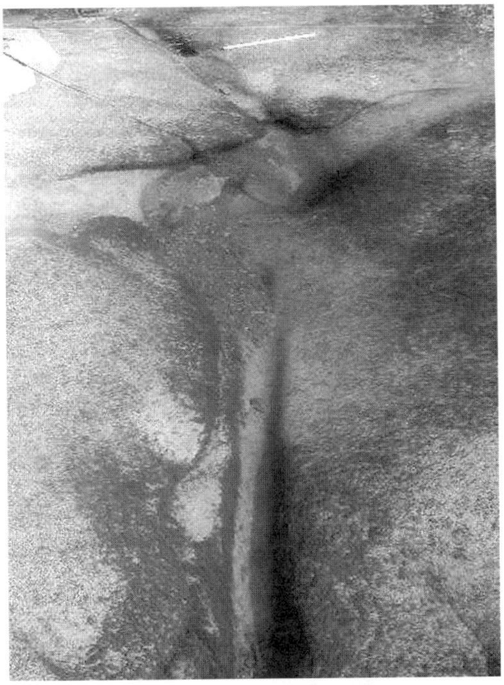

Figure 47. *Straight parallel-sided furrow* (lower half of photo). River Lune (Halton), UK. Flow from top to bottom. Fine-grained sandstone. The scale is 60 cm long.

(ii) *Curved parallel-sided furrow*: *Parallel-sided furrows* are often found to curve substantially and consistently to one side (Fig. 48). The angle of curvature ranges from 0° to 90°. They may begin with a longitudinal orientation at their upstream end and curve away from this, or curve toward a longitudinal orientation with distance downstream, but overall the dominant orientation is longitudinal. Both *straight* and *curved parallel-sided furrows* may originate at their upstream ends at some defect (e.g., Tinkler and Stenson, 1992; Pollard et al., 1996) (Fig. 49), in which case they may be regarded as very long *flutes*, or they may be unrelated to any obvious defect.

(iii) *Sinuous parallel-sided furrow*: A *parallel-sided furrow* that contains curves of both senses is termed *sinuous* (Fig. 50). Wohl and Achyuthan (2002) describe such features in an unnamed river in India, while Bryant (2001: p. 85) describes *sinuous parallel-sided furrows* in coastal areas formed by tsunamis.

(iv) *Cuspate* and *noncuspate parallel-sided furrows*: As with *short furrows*, the rims of *parallel-sided furrows* may in general be cuspate or rounded.

(v) *Parallel-sided furrow with levees*: This is a rare type of *furrow*, only two examples of which were observed at a single site (Sichiri-gawa). They were adjacent and occurred in a marginal area of the channel that experienced frequent wetting and drying cycles, in a lithology prone to flaking, and consisted of a *furrow* with rounded margins that were raised above the general level of the surrounding bedrock surface (Fig. 51). It is thought that these *levees* are caused by a reduced rate of erosion adjacent to the *furrow*, where the rock does not dry out as often as a result of water ponded within the *furrow*, thus inhibiting the flaking process (H. Ikeda, 1999, personal commun.).

Figure 48. *Curved parallel-sided furrow*. Nam Mae Chaem (Ob Luang), Thailand. Upstream to top right of picture. Granitic gneiss. The notebook (arrowed) is 15 cm long.

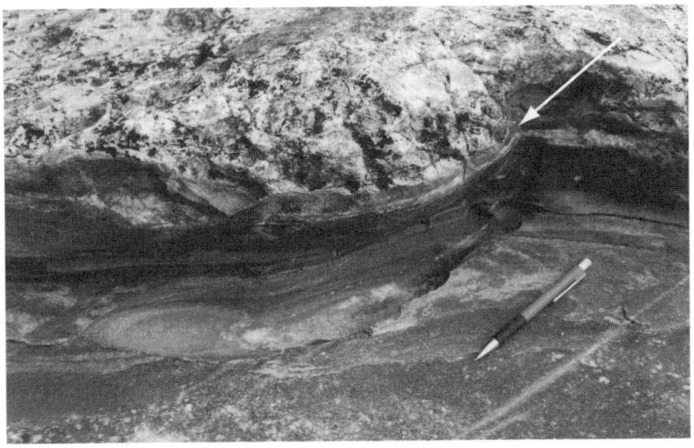

Figure 49. *Parallel-sided furrow* originating at a "defect," which in this case is an internal corner formed at a joint (arrowed). Nam Mae Khlang, Thailand. Flow from right to left. Gneiss. Pen for scale.

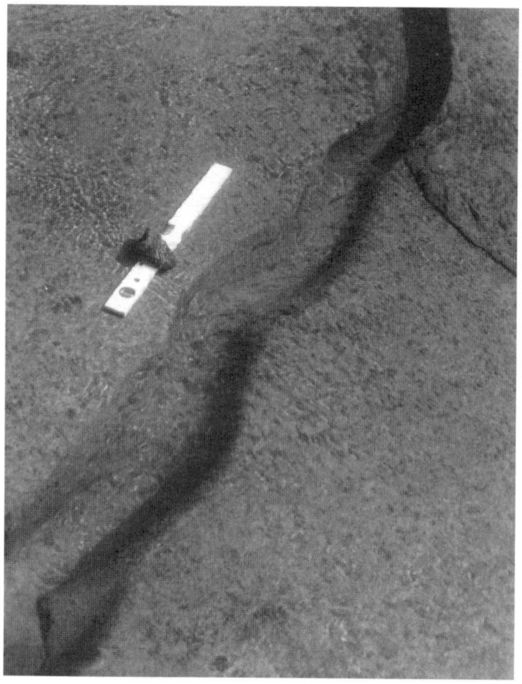

Figure 50. *Sinuous parallel-sided furrow*. River Lune (Halton), UK. Flow from top right to bottom left. Fine-grained sandstone. The scale is 60 cm long.

(vi) *Shear zone furrow*: This is also a rare type of *furrow*, first described by Tinkler (1997a), which he referred to as a "meandering channel" at a "flow convergence zone" in Fifteen Mile Creek, Ontario. It requires special circumstances for its formation, namely a large obstacle to the flow that creates a wake area of slower flow, and a shear zone between this and the main flow. The *furrow* occurs on the channel floor within the shear zone and has rounded rims. Tinkler described an extensive example of such a feature, 14 m long, *sinuous* with *levees*, and punctuated by potholes along its length. He attributed it to vertical vortices shed from the obstacle and advecting within the shear zone, eroding the *furrow* where they impinge on the bed. The example illustrated in Figure 52 is much smaller and simpler, but apparently similar in origin.

(vii) *Chute furrow*: Immediately above waterfalls and *steps* (section 6.2c), and around the edges of headcuts, where flow is rapid but is still in contact with the rock surface, the water may carve out straight, steeply dipping, *parallel-sided furrows* (Fig. 53). Where the water flows over a wide lip before going into a fall, many such *chute furrows* may occur, sometimes in a conjugate fashion. The flow in these bedforms is clearly supercritical, hence the adjective *chute*. *Chute furrows* may become deeply incised and slot-like, with thin and delicate ridges of rock dividing adjacent individuals. The *furrows* described as "sweep flutes" by Maxson and Campbell (1935) were found in areas of rapid

Figure 51. *Parallel-sided furrow with levees*. Sichiri-gawa, Japan. Flow from bottom to top. Fine-grained sandstone. Pen for scale (arrowed).

Figure 52. *Shear zone furrow* (arrowed). Note the obstacle to flow (boulder) and the position of the *furrow* downstream of the corner of it. Birk Beck, UK. Flow from top to bottom. Medium-grained sandstone. The scale is ~30 cm long.

Figure 53. *Chute furrow* at the lip of a headcut. Lower Nahal Zin, Israel. Limestone.

flow, such as the lips of waterfalls, and may have been *chute furrows*.

(viii) *Chimney furrow*: As the slope of a *chute furrow* increases, a point is reached at which the *furrow* is vertical or near-vertical, and at which water within the *furrow*, though remaining in contact with the rock surface, is essentially in free fall. Such a feature is here named a *chimney furrow*, and examples are found in the faces of waterfalls and headcuts (Fig. 54A). Some *chimney furrows* are even overhanging. The cross sections of *chimney furrows* range from gently concave through semicircular to deeply incised notches. Some such incised *chimney furrows* have a "keyhole" type of cross section (Fig. 54B). In addition to the continuum in slope between *chute furrows* and *chimney furrows*, some bedforms exist as *chute furrows* at their upstream ends and pass downwards into *chimney furrows*.

(ix) *Bifurcating furrow*: a *straight* or *sinuous furrow* that divides into two branches in a downstream direction (Fig. 55). The upstream-facing surface between the two arms at the dividing point may take the form of a raised, bulbous knob, or it may be low and gently rounded. In the former case, it takes on the appearance and properties of an *obstacle* with associated *current crescent* (section 3.1a). King (1927) describes *bifurcating furrows* in Bar-

Figure 54. A: An assemblage of *chimney furrows* in a waterfall (dry at time of photograph). The slope of the *chimneys* ranges from nearly vertical to slightly overhanging. Nahal Hatira, Israel, looking upstream. Interbedded limestone and marl. Field of view ~20 m across in center of picture. B: View vertically down a *chimney* with a "keyhole" cross section. Nahal Hatira, Israel. Interbedded limestone and marl. The circular region of the bedform in the center of the picture has a diameter of approximately 1.5 m.

Figure 55. *Bifurcating furrow*. Woolshed Creek, Australia. Flow from top right to bottom left. Microgranite. The length of the *furrows* from the bifurcation to the edge of the photo is ~3 m.

ton Creek, Texas, and Shaw et al. (2000) describe large-scale (tens of kilometers long) bifurcating linear depressions that purport to be eroded into till by subglacial meltwater in Alberta, Canada. The fields of en echelon sichelwannen (s-forms) illustrated by Kor et al. (1991) and Shaw (1996) could also be considered analogous to groups of interconnected *bifurcating furrows*.

(x) *Furrow with regularly spaced depressions*: a feature described and named by Wohl (1993) from Piccaninny Creek, Australia. It consists of a shallow, linear *furrow* containing regularly spaced short, elliptical depressions orientated obliquely to the *furrow* (Wohl, 1993: Fig. 4 therein). It forms the initial stage of a sequence of bedforms that progresses with the deepening of the *furrow* and of the depressions within it, eventually forming an *inner channel with wall undulations* (see also sections 1.2f and 6.1c).

(xi) *Group of parallel-sided furrows*: Several authors have described *groups of parallel-sided furrows* from bedrock (e.g., King, 1927; Blank, 1958; Baker, 1973; Shepherd and Schumm, 1974; Wohl, 1993), experimental dissolution studies (Allen, 1971a), and sedimentary environments (Dzulynski and Sanders, 1962; Karcz, 1967, 1973; Allen, 1969; Ikeda, 1978), some reporting them to be extensive and to have regular transverse spacing (e.g., Blank, 1958; Allen, 1969, 1971a; Baker, 1973; Ikeda, 1978; Wohl, 1993). Those of Ikeda (1978) were found to scale with bankfull depth, while those of Allen (1969, 1971a) were on a much smaller scale and had a transverse spacing close to that found by Kline et al. (1967) for low-speed streaks within the viscous sublayer. It would appear that the extensive, regularly spaced variety, in which individuals are typically tens or hundreds of meters long, require wide, uniform flows over *plane beds*. Groups of such *furrows* have also been observed on a high-energy intertidal rock platform in an estuarine environment by Allen (1987). Meanwhile, other less extensive and more irregular *groups of parallel-sided furrows* may occur on channel margins or immediately above *knickpoints* (Fig. 56).

(d) *Compound parallel-sided furrows*. It is often found that *parallel-sided furrows* have undulating walls and floors and consist of not a single depression, but two or more closed depressions divided by cols, which may be rounded or cuspate. These intervening highs in the floor of the *furrow* are significantly lower than the rim of the *furrow* as a whole. Such bedforms are named *compound parallel-sided furrows*. It is possible that there is a maximum "stable" length (relative to the width) for simple *parallel-sided furrows* and that beyond this length they become divided internally into component *furrows* in a compound bedform, but this is speculative. See also section 3.2b for descriptions of further linear features with undulating floors. There are two basic types of *compound parallel-sided furrows*.

(i) *Regular compound furrow*: This takes the form of a linear series of similar *furrows* (either *short* or *parallel-sided*), which are contiguous; they may abut end to end, or the start of each may overlap slightly with its upstream neighbor, being laterally offset from it in a stepwise fashion. These component *furrows* are typically separated by sharp crests and together form a long *compound paral-*

Figure 56. *Group of parallel-sided furrows* immediately above a *knickpoint*. Wadi Hazazon, Israel, looking upstream. Limestone. Camera bag, 20 cm wide, for scale.

lel-sided furrow of repetitive morphology (Fig. 57). The crests separating the component *furrows* may be laterally continuous with *sharp-crested hummocky forms* (section 2.1d), if present, on the bedrock surface on either side of the *compound furrow*. Sometimes, a *regular compound furrow* is formed from a series of *potholes* in which the exit furrow from one *pothole* leads directly into the next *pothole*. Rarely, *compound furrows* are not linear, but form a zigzag pattern in plan view, with each component *furrow* meeting its neighbor at a high angle. This occurs in narrow, sinuous channels and appears to be a result of the deflection of flow from alternate channel walls at bends. The zigzag pattern of the *compound furrow* follows the channel sinuousity. It is also possible for *regular compound furrows* to be curved, so that at one end they are longitudinal, while at the other they are transverse. The very long *furrows* described by Blank (1958: Fig. 4 therein) from the James River, Texas, and that illustrated by Sevon (1988: Fig. 4 therein) from the Susquehanna River, Pennsylvania, may be assigned to the category of *regular compound furrows*. In Sichiri-gawa (Table 2), stepwise *regular compound furrows* dominate the morphology in the part of the reach parallel to strike, while in the River Dee (Table 2), an *inner channel* can be seen to possess the same type of morphology.

(ii) *Irregular compound furrow*: Whereas *regular compound furrows* contain a series of similar depressions, the *irregular* variety contains a series of randomly spaced depressions that are dissimilar in type and/or size. The internal morphology of *irregular compound furrows* does not repeat itself and may be complex (Fig. 58). In planform, they may be straight or sinuous. Such *furrows* also occur as a type of s-form (Kor et al., 1991; Shaw, 1996).

(e) *Expanding furrows*. These *longitudinal furrows* are not parallel-sided, but rather they become wider in the downstream direction.

Figure 58. *Irregular compound parallel-sided furrows*. The dashed lines run along the center of two *compound furrows*. Nahal Zin, Israel, looking upstream. Limestone. Camera bag, 20 cm wide, for scale.

Figure 57. *Regular compound parallel-sided furrow*. Than Rattana, Thailand. Flow from top to bottom. Andesite. The notebook is 15 cm long.

Figure 59. *Funnel-shaped furrow* (underwater). Stakeley Beck, UK. Flow top right to bottom left. Medium-grained sandstone. The scale is 60 cm long.

(i) *Funnel-shaped furrow*: a *furrow* that expands in a uniform manner with distance downstream (Fig. 59).

(ii) *Bulbous furrow*: a *furrow* that is parallel-sided for some distance but then expands rapidly into a deeply scoured, wider hollow (Fig. 60). Based on the presence of sharp crests between the narrow and wide parts of the bedform, it is predicted that flow enters the wider part from the parallel-sided part as a jet and generates counterrotating gyres in the two halves of the wider portion on either side of the jet (see Discussion for inferences to be drawn from sharp crests).

(f) *Runnels* (in general) and *runnels with wall relief*. (See also section 3.2b.) *Runnels* are important conduits for flow at low stage, when they convey a significant proportion of the channel's discharge. *Runnels* are a special type of *furrow* and may assume many of the morphologies described above. However, some remarks specifically on *runnels* are necessary here. Even if they are not structurally influenced, *runnels* are often not parallel to the main channel and may be sinuous, because they follow the small-scale topography of the channel floor, connecting its lowest points. At a stage when most of the channel is dry or consists of discontinuous shallow bodies of water, *runnels* preserve within themselves rapid, typically transcritical flow. Other sculpted forms, such as *sharp-crested hummocky forms* (section 2.1d) and *parallel-sided* and *short longitudinal furrows*, may occur within *runnels* (see, for example, section 3.2b). These forms, be they transverse or longitudinal, are often orientated according to flow along the *runnel* they are in, rather than to flow parallel to the main channel (if different), and follow any changes in the direction of the *runnel* relative to the main channel with distance downstream. This observation demonstrates that low-stage flow within the *runnel* is responsible for the generation of sculpted forms, and that it remains geomorphologically effective, possibly by maintaining a transcritical state (see Discussion). Many *runnels* may be found that possess pronounced wall undulations, and here these are termed *runnels with wall relief*. The wall topography is certainly hydrodynamic and not structural in origin, as it is periodic in nature and occurs in homogeneous substrates. Two types of *runnels with wall relief* have been identified:

(i) *Runnel with cusped margins*: Only one example of such a *runnel* was observed by the authors (Fig. 61). The relief and wavelength of the cusps, and the depth of the *runnel*, are a small fraction of the *runnel*'s width. The cusps resemble a rounded sawtooth pattern and point downstream. The floor of the *runnel* is not without topography either; each indentation in each margin of the *runnel* is associated with a small, rounded depression in the bed adjacent to it, but shifted slightly downstream from it.

(ii) *Runnel with alternating scour*: This is a morphology that appears to be common in relatively soft but cohesive substrates. For instance, it was developed experimentally by Wohl and Ikeda (1997) using a bentonite-sand mixture, and by Kodama and Ikeda (1984) using silty clay. The authors have also seen it in unconsolidated Plio-Pleistocene sediments at Byōbuga Ura, Japan (Fig. 62). However, this bedform was first described from sandstone in Piccaninny Creek, Australia, by Wohl (1993), where it forms part of a downstream sequence of bedforms. The sequence begins with a *furrow with regularly spaced depressions* (section 1.2c), which becomes progressively deeper and

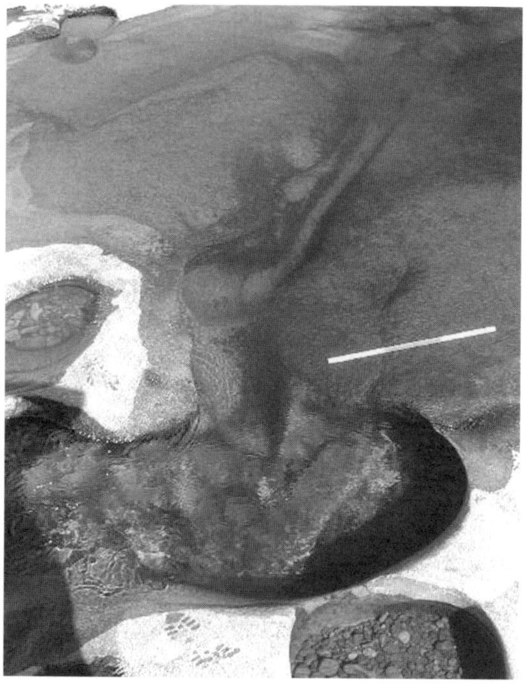

Figure 60. *Bulbous furrow* (underwater). The linear part of the *furrow*, approaching the center of the photograph from top right, expands rapidly into a near-circular depression. River Lune (Halton), UK. Flow from top to bottom. Fine-grained sandstone. Scale is 60 cm long.

Figure 61. *Runnel with cusped margins*. Sichiri-gawa, Japan. Flow from right to left. Fine-grained sandstone. Lens cap for scale.

Figure 62. *Runnel with alternating scour.* Note that depressions with bends of different senses are present (see text for explanation). The standing figure is Hiroshi Ikeda. Byōbuga Ura, Chiba Prefecture, Japan. Flow from top to bottom. Unconsolidated Plio-Pleistocene sediments.

more sinuous with distance downstream, as the depressions grow into offset *lateral potholes* lining each side of the *furrow*. This is the part of the sequence that is here called *runnel with alternating scour*. The sequence culminates in the coalescence of the *potholes*, forming an *inner channel with undulating walls* (section 6.1c). A high gradient is required to produce this type of *runnel*. Wohl and Ikeda (1997) report that it was developed at a flume gradient of 5%–10%, and Wohl (1993) reports a gradient of 4% for the *furrow with regularly spaced depressions* that is a precursor to the *runnel with alternating scour*.

The morphology of a *runnel with alternating scour* is complex. The principle feature is a sinuous thalweg punctuated by a series of scour hollows, which are typically on alternate sides of the *runnel* and are formed at the bends in the thalweg, as the locus of maximum erosion weaves from one side of the *runnel* to the other. The hollows range from slight depressions with a widening of the thalweg, to *lateral potholes* where they are most fully developed. The scour hollows and sinuous thalweg are associated with a *step-pool*–type longitudinal profile, the steps occurring between successive hollows. At the foot of each step, the thalweg plunges into a scour hollow and, at the same time, turns sharply so that it emerges from the hollow and crosses to the opposite side of the *runnel*. Here it turns again as it goes down the next step and plunges in to the next hollow. In this form, the hollows are on alternate sides of the *runnel* and contain turns of the thalweg of opposite senses (Fig. 62). An alternative form exists in which the hollows are all on the same side and contain turns of the same sense. In this case, the thalweg crosses the *runnel* as it emerges from a hollow and crosses back again as it goes down the step. This form is illustrated by Wohl (1993: Fig. 4B therein). While *step-pool* systems, which occupy the full width of the channel, share the stepped longitudinal profile of *runnels with alternating scour*, they differ not only in scale, but also in their lack of a sinuous thalweg with regularly repeating wall relief.

1.3 Nonlongitudinal Furrows

Nonlongitudinal furrows are so named either because they are not parallel to the local flow direction or because this direction deviates greatly from the general direction of the channel. *Furrows* in this class are all interpreted as containing reversed flow to a greater or lesser extent at formative discharges. These *furrows* form a series of bedforms with increasing degrees of deviation from the orientation of the channel (Fig. 63).

(a) *Oblique sloping furrows.* A common feature of bedrock channels with margins of moderate slope (between ~10° and 40°) is *parallel-sided furrows* that slope down the channel margin at an oblique angle (Fig. 64). These *furrows* may occur in groups of parallel individuals. The rims may be cuspate or rounded. The flow structures contained within these *furrows* are at present a matter of conjecture, but it is hypothesized here that the *furrows* are orientated obliquely to the local flow direction, which is thought to be similar to the direction of the channel, and that the rim nearest the thalweg acts as a defect, causing flow separation. The resulting flow separation cell consists of strongly helical flow along and within the *furrow*, similar to that described by Allen (1971a) for flow adjacent to the rims of *flutes*. As the dip of the channel margin and of the *furrow* increases toward the vertical, and/or the orientation of the *furrow* is more directly downslope toward the channel center, *oblique sloping furrows* appear to grade into features more appropriately termed *linear lateral potholes* and *transverse furrows* respectively. *Oblique sloping furrows* are not subdivided.

(b) *Transverse furrows.* *Furrows* orientated approximately transversely to the local flow direction are much less common and diverse than their longitudinal counterparts. They appear in general to be the result of (1) a flow separation cell (a roller) in the lee of a sharp convex change of slope (i.e., a positive topographic "defect") with considerable transverse extent and (2) enhanced scour in the region of reattachment. They are therefore equivalent to Ängeby's (1951) "leeside potholes." The defect, however, is not always apparent. In such cases, it may have been eroded away, and the *transverse furrow*, once established, becomes self-sustaining by generating its own flow separation cell. However, some *transverse furrows* exist on surfaces lacking obvious defects and have low relief and rounded rims, and are therefore not sufficiently well developed to be explained in this way (see Discussion for guidelines on inferred flow patterns within bedforms). It therefore appears that some, as yet unidentified, mechanism exists by which *transverse furrows* can be generated in the absence of a defect and its associated roller. *Transverse furrows* differ from the other groups of *furrows* described in this typology in being generally rather irregularly shaped and spatially poorly defined in comparison. As with many bedforms in bedrock channels, *transverse furrows* are not restricted to the horizontal, but as the orientation of the *furrow* approaches the vertical, it is best termed a *linear lateral pothole*. Jennings (1983) describes *transverse furrows* from the Snowy River, New South Wales, although some of the features he describes in this category would be better classified as *obstacle marks* (section 3.1). The "transverse erosional ripples" of Shepherd and Schumm (1974: Fig. 4 therein) appear in fact to be *transverse furrows* within an otherwise planar surface. Bryant (2001: p. 89) described large-scale (up to 50 m long) *transverse furrows* due to tsunamis on the lee sides of positive topographic defects in coastal environments. *Transverse furrows* are also known to occur in subglacial environments and are included as a type of s-form (Kor et al., 1991). Ancient *transverse furrows* in bedrock (i.e., preserved in unconformities) have been described by Shrock (1948: p. 230–236). Features similar to *transverse furrows* also occur as a result of flow separation upstream of an obstacle to the flow, but these are considered as a type of *obstacle mark* (section 3.1a). Three types of *transverse furrows* are identified:

(i) *Simple transverse furrows*: These are *furrows* of simple morphology consisting of a single depression (Fig. 65). They may be straight or slightly curved. Sevon (1988: Fig. 6 therein) illustrates a *simple transverse furrow* in the lee of a ridge. The "undercut downstep" of Tinkler (1993) is also a type of *simple transverse furrow*, the defect in this case being the *step*.

(ii) *Compound transverse furrows*. These are *transverse furrows* with more complex morphology, consisting of a series of small, closed depressions linked in a transverse direction, and undulating sidewalls (Fig. 66). Their morphology is always of the *irregular compound* type, i.e., they are composed of dissimilar, nonrepeating, and irregularly spaced component *furrows* and *potholes*.

(iii) *Cross-channel furrow*. This is a bedform first described by Tinkler and Wohl (1998b: Fig. 7 therein). It consists of

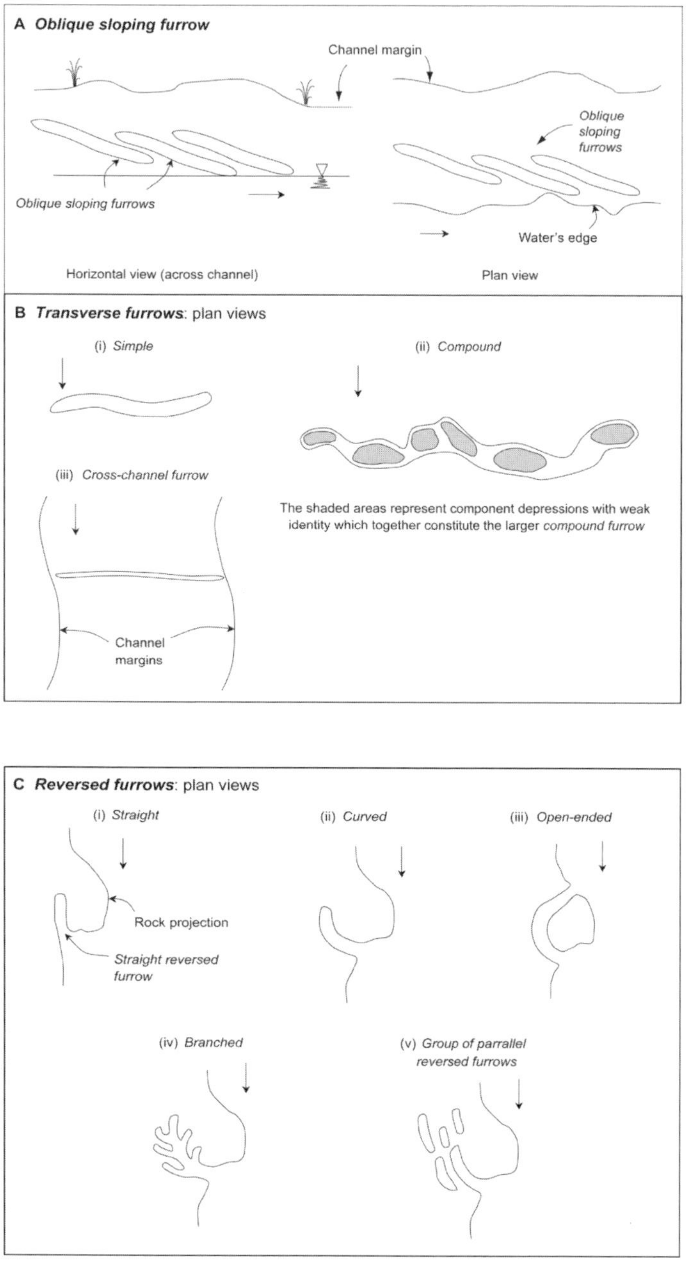

Figure 63. *Nonlongitudinal furrows*. Arrows indicate flow directions (where appropriate).

Figure 64. *Oblique sloping furrows*. Two examples (arrowed) can be seen to the left of the notebook. Nam Mae Chaem (Ob Luang), Thailand. Flow from top left to bottom right. Granitic gneiss. The notebook is 15 cm long.

a *transverse furrow* with a relatively uniform, U-shaped cross section, cuspate rims, and a length of approximately the channel width (Fig. 67). The initial defect, if ever present, is no longer apparent in these cases.

(c) *Reversed furrows.* *Furrows* may often be found in the lee of large rock projections and irregularities, either alone or forming part of a group. Probably for reasons relating to the proportion of time for which they are exposed, these *furrows* are most often seen in steeply sloping channel margins but also occur on horizontal surfaces in the channel floor. These *furrows* are variable in orientation and may be strongly curved but are typically orientated subparallel to the main channel. They are open at the downstream (with respect to flow in the main channel) end, which is in the lee of the projection, and generally taper and close toward the upstream end. At the upstream end, these *reversed furrows* may terminate at a steep and high headwall, indicating that the *furrows* are deeply incised into the projection at this end. In such cases in steeply sloping surfaces, the headwall and adjacent *furrow* walls create a depression similar to a *lateral pothole*. *Reversed furrows* are interpreted as being eroded by the recirculating flow in the lee of the projection, and therefore as containing upstream-directed flow. In this sense, they differ fundamentally from both *oblique sloping furrows* and *transverse furrows*, for whereas in the latter types the local flow is assumed to be subparallel to the general direction of the channel, and the *furrows* deviate from this orientation, *reversed furrows* are parallel to the local flow, which is a separated, upstream-directed flow. *Reversed furrows* may be steeply dipping, but in common with *oblique sloping furrows*, as they approach the vertical, their hydrodynamics must change from that just described to that appropriate to *lateral potholes*, and they should be classified accordingly. *Reversed furrows* may also become very wide or strongly tapered so that they are no longer a linear feature, and

Figure 65. *Simple transverse furrow* (underwater). Flow from left to right. River Lune (Halton), UK. Fine-grained sandstone. The scale is 60 cm long.

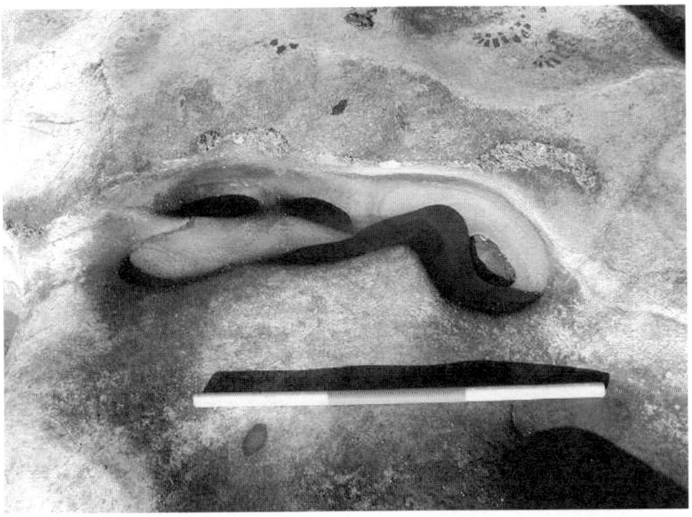

Figure 66. *Compound transverse furrow*. Note that the bedform contains several centers of erosion. River Lune (Halton), UK. Flow from top to bottom. Fine-grained sandstone. The scale is 60 cm long.

Figure 67. *Cross-channel furrow* (underwater). Sichiri-gawa, Japan. Flow from left to right. Fine-grained sandstone. Field of view at water's edge is ~4 m.

they would likewise be more correctly referred to as a *lateral pothole* for this reason. Several types of *reversed furrows* exist:

(i) *Straight* (Fig. 68).

(ii) *Curved*: *Reversed furrows* will often curve with a consistent sense toward the channel center (Fig. 69).

(iii) *Open-ended*: In steeply sloping marginal areas, if the *furrows* are long and sufficiently curved, they may actually pass right around the projection on the side away from the channel and rejoin the main channel upstream of the projection (Fig. 70). Such *furrows* are then open at both ends.

(iv) *Branched*: This type of *reversed furrow* is composed of a main, curved *furrow* with one or more arms branching off tangentially from it (Figs. 69 and 71).

(v) *Group of parallel reversed furrows*: Several *reversed furrows* may occur as a group of neighboring, but discrete, parallel individuals, either in series or alongside each other. They are all fed by the same eddy (Fig. 72).

1.4 Furrow Complexes

Furrow complexes are a group of special *compound furrows*, in which two or more component *furrows* with shared boundaries form a distinctive geometrical arrangement (Fig. 73). This suggests that the hydrodynamics of the component *furrows* are closely related. Four such distinctive arrangements are identified:

(a) *Convergent.* Only one example of this type of complex was observed by the authors. It consists of three *oblique sloping furrows* whose axes are convergent downstream, and which merge downstream into a single, broad *furrow* (Fig. 74).

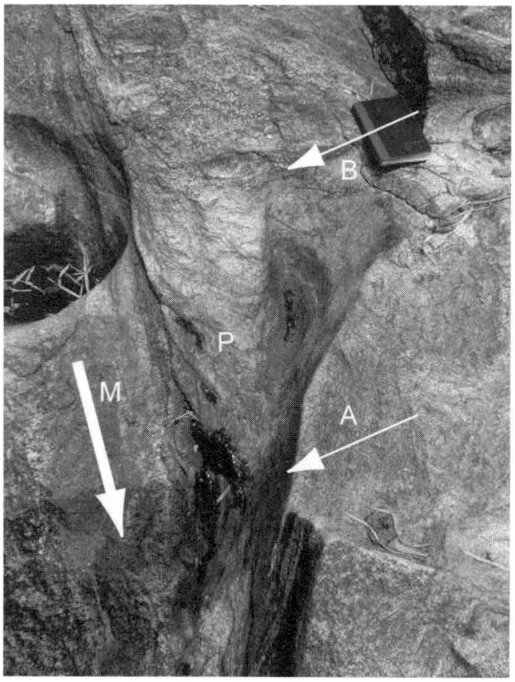

Figure 68. *Straight reversed furrow*, seen from above. Flow in the main channel flows in the direction shown by the arrow M. A *straight reversed furrow*, with its ends shown by arrows A and B, is interpreted as having formed in response to a flow recirculation cell in the lee of a rock projection (P). Nam Mae Chaem (Ob Luang), Thailand. Granitic gneiss. The notebook is 15 cm long.

Figure 69. *Curved reversed furrow*. This example also has two *branches* (B). Nam Mae Chaem (Ob Luang), Thailand. Upstream to top. Granitic gneiss. Pen for scale.

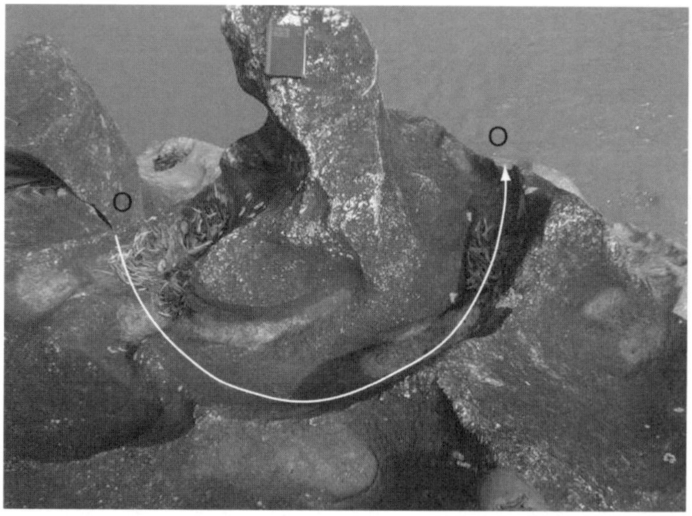

Figure 70. *Open-ended reversed furrow*. The inferred flow direction within the *furrow* is shown by the arrow, and the open ends of the *furrow* are indicated by the letter O. Nam Mae Chaem (Ob Luang), Thailand. Upstream to right. Granitic gneiss. The notebook is 15 cm long.

(b) *Yin-yang.* This "head-to-tail" arrangement of two contiguous *furrows* was first identified by Whipple et al. (2000a). These authors described the small-scale secondary internal sculpting of *potholes* by *curved short furrows* (section 1.1c) that exhibited this relationship. However, *yin-yang* complexes can exist on a larger scale in *furrows* on any surface, as shown in Figure 75.

(c) *Bifurcating.* The *furrows* trend parallel to flow but both join and divide as they do so (Fig. 76). The "en echelon" *sinuous furrows* described by Bryant (2001: p. 85) would be better described as *bifurcating furrow complexes*.

(d) *Nested curved.* This consists of *curved parallel-sided furrows* with curvature of the same sense lying side by side (Fig. 77).

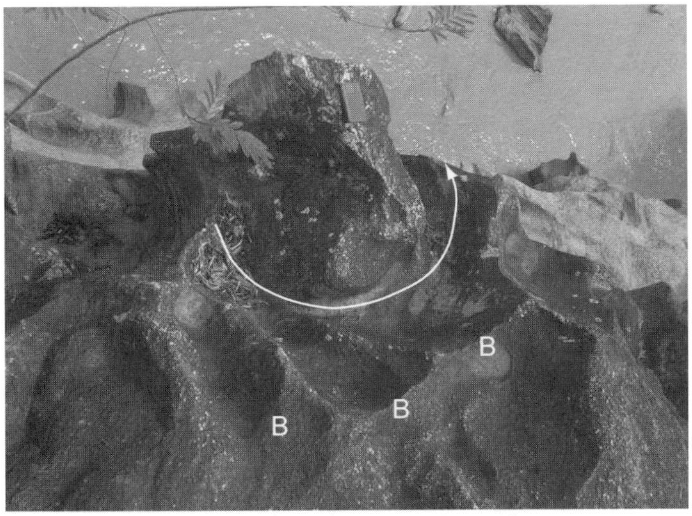

Figure 71. *Branched reversed furrow*. The main part of the *furrow* and the inferred flow direction within it are shown by the arrow, and each of its three branches is indicated by the letter B. Nam Mae Chaem (Ob Luang), Thailand. Upstream to right. Granitic gneiss. The notebook is 15 cm long.

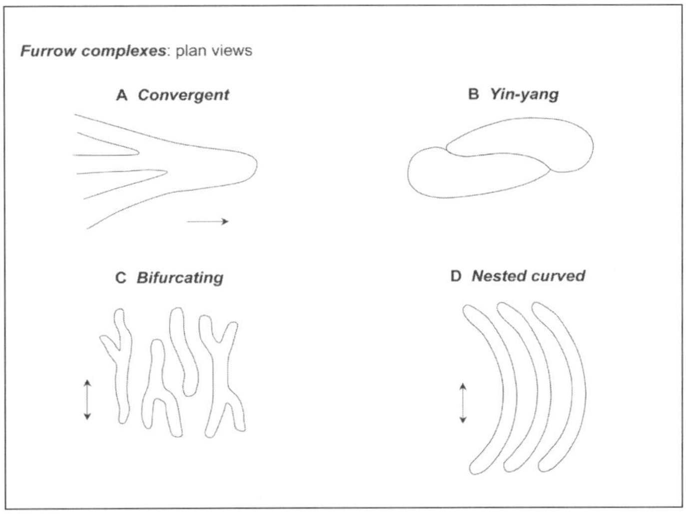

Figure 73. *Furrow complexes*. Arrows indicate flow directions.

Figure 72. *Group of parallel reversed furrows*. Three *parallel reversed furrows* can be seen in the group, each indicated by the letter F. The group has formed in response to the rock projection P and the flow separation immediately downstream of it (white arrow). Nam Mae Chaem (Ob Luang), Thailand. Upstream to right. Granitic gneiss. The notebook is 15 cm long.

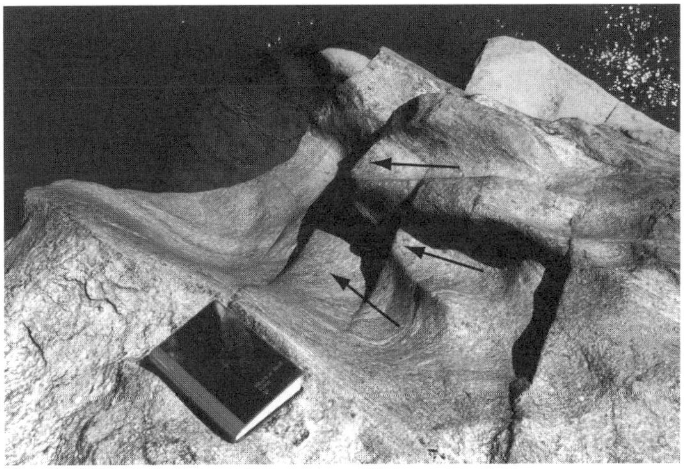

Figure 74. *Convergent furrow complex*. The convergence is indicated by arrows. Nam Mae Chaem (Ob Luang), Thailand. Flow from right to left. Granitic gneiss. The notebook is 15 cm long.

Figure 75. *Yin-yang furrow complex* (dashed lines). River Lune (Tebay), UK. Upstream to left. Calcareous mudstone. The scale is 60 cm long.

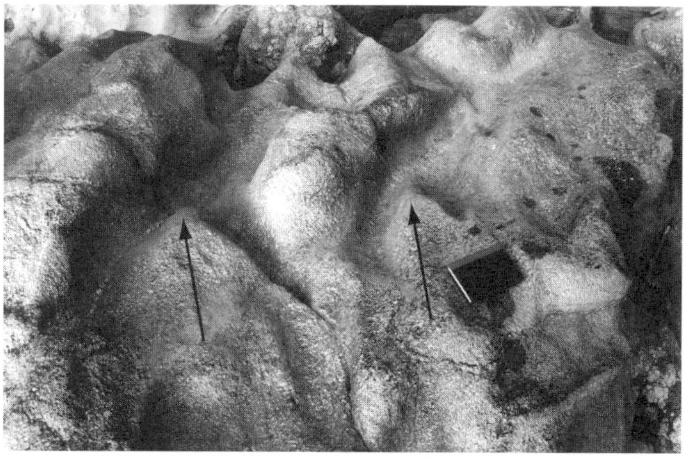

Figure 76. *Bifurcating furrow complex*. Bifurcation points are indicated by arrows. Nam Mae Chaem (Ob Luang), Thailand. Upstream at bottom. Granitic gneiss. The notebook is 15 cm long.

1.5 Overhanging Concave Features

This is a class of *concave features* that occur in a vertical or near-vertical face and that undercut the face (Fig. 78). They range from deep sculpted forms with an approximately horizontal roof to those of lower relief in which the back wall arches over but does not approach the horizontal. Although some *lateral potholes* have an overhanging back wall, they are distinguished from the bedforms below by the fact that the overhang is not great, and moreover is incidental, rather than an essential feature of the bedform. *Overhanging concave features*, on the other hand, are defined by their pronounced overhang, which is their main feature.

(a) *Overhanging concave surface.* In gorges and channels with vertical margins, there are often found smooth, *overhanging*

Figure 77. *Nested curved furrow complex* (partly underwater). The ends of four *furrows* within the complex are indicated by arrows. Birk Beck, UK. Flow from bottom to top. Medium-grained sandstone. Pencil for scale.

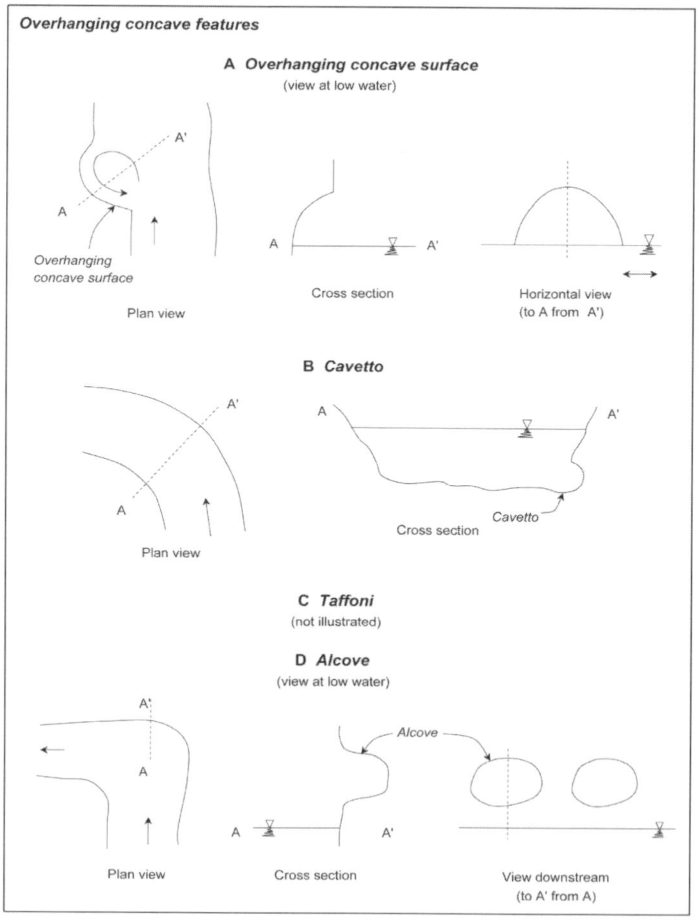

Figure 78. *Overhanging concave features.* Arrows indicate flow directions (where appropriate). Dashed lines indicate lines of cross sections.

concave surfaces, either alone or in conjugate form (Fig. 79). They are variable in size, which may be comparable to the width of the channel. These surfaces occur within flow recirculation cells in the lee of projections or at abrupt channel expansions and may face either across the channel or downstream. The upstream rim of the surface, where flow separation occurs, is typically sharp. Sato et al. (1987: Fig. 20 therein) considered such surfaces to be large *breached potholes* and termed them "broken giant kettles" even though this supposed origin is not clear in their examples. Lorenc et al. (1994: Fig. 6D therein) also illustrate *overhanging concave surfaces*.

(b) *Cavetto.* Erosion is naturally most intense on the outside of a bend, and in some cases this erosion becomes focused at a particular level, forming a *cavetto* (Fig. 80). A *cavetto* is essentially a *parallel-sided furrow* on the outside of a bend, eroded into a near-vertical face so that the upper rim of the *furrow* is strongly overhanging. *Cavettos* vary greatly in size, and in deep, narrow channels their vertical dimension may approach the width of the channel. In the case of dip-parallel reaches, they are often structurally influenced, tending to pick out less resistant beds. Bryant (2001: p. 88) described *cavettos* from coastal environments due to tsunamis. *Cavettos* also occur in subglacial environments as a type of s-form (Kor et al., 1991; Kor and Cowell, 1998).

(c) *Taffoni (tafoni)*: This is a cavernous weathering landform most commonly found in acid to intermediate plutonic rocks but also found in sandstone, limestone, and schist, and which occurs in semiarid climates (Fairbridge, 1968: p. 1103). It is generally attributed to the mechanical disintegration of the rock due to its impregnation by salts and the crystallization of these salts during repeated wetting and drying cycles (Sparks, 1972: p. 28–29, 320, 345). *Taffoni* can be found on the upper reaches of the walls of gorges as an open boxwork structure (Fig. 81). However, as one descends toward the floor of the gorge, the hollows lose definition, presumably owing to the increasing degree of fluvial erosion. *Taffoni* range in size from a few centimeters to ~2 m (Sparks, 1972: p. 29). Small examples are sometimes called "alvéoles."

(d) *Alcove.* *Alcoves* are cave-like hollows with a near-horizontal floor that occur in the vertical walls of gorges and *inner channels* (Fig. 82). They appear to be related to *cavettos* in that they occur in association with them on the outside of a bend.

Figure 79. *Overhanging concave surface.* Nam Mae Chaem (Ob Luang), Thailand. Flow from left to right. Granitic gneiss. Field of view is ~3 m.

Figure 80. *Cavetto.* Nahal Ashalim, Israel, looking downstream. Limestone. Sunglasses (arrowed) for scale.

Figure 81. *Taffoni.* Nahal Shani, Israel, looking upstream. Fine-grained sandstone. The largest hole is ~30 cm across.

In fact, they occur on sharp bends, such that the *alcoves* face upstream toward the reach above the bend. Several *alcoves* may occur adjacent to each other. They may also exhibit a hierarchical structure, with small *alcoves* forming in the walls of large ones. It is possible that they originate as overdeepened points within a *cavetto*, where erosion is particularly intense or the rock particularly weak, these points being unstable and developing by positive feedback into a deep hollow.

1.6 Shallow Concave Surfaces

These forms do not fit neatly into any of the above categories, and so are placed in a class of their own. They are not subdivided. *Shallow concave surfaces* observed by the authors are up to about a meter in length and vary from being elongated parallel to flow to equidimensional. They have a low curvature, their relief being a small fraction of their length (Fig. 83). Their rims may be sharp or rounded. Their affinities and associated flow structures are uncertain, except that flow separation must occur at the boundaries of sharp-rimmed varieties (see Discussion). As their relief increases, they form a continuum with *lateral potholes*. *Shallow concave surfaces* can be seen in the illustrations of Thybony (2000).

Figure 83. *Shallow concave surfaces* (arrowed). River Lune (Tebay), UK. Flow from top left to bottom right. Calcareous mudstone. The hammer is 30 cm long.

2. Convex and Undulating Surfaces

This group contains a much smaller variety of forms than the *concave features*. It is in general composed of various smoothed and rounded surfaces, although some bedforms have sharp crests. Whereas the areal extent of *concave features* is well defined, especially in sharp-rimmed varieties, that of *convex* and *undulating surfaces* is poorly so. *Convex surfaces* arise in three main ways: (1) through the abrasion and rounding of angular bedrock highs of structural origin created by plucking; (2) as relict highs between *concave features* in which erosion rates are much higher; and (3) as relict highs where erosion rates are low because the rock is more resistant. *Undulating surfaces* are evidence of spatially fluctuating erosion rates; this may occur either in an irregular fashion or in a quasi-regular, cyclical fashion.

Figure 82. *Alcoves*. Nahal Neqarot, Israel, looking downstream. Limestone. Person for scale.

2.1 Hummocky Forms

Areas of smooth, undulating bedrock without distinct *furrows* are common in bedrock channels, especially in limestone, and these are termed *hummocky forms* (Fig. 84). Both rounded and sharp-crested varieties occur. While they do develop at a characteristic scale at any particular site and are always on the decimeter scale, they cannot in general be described as periodic. *Hummocky forms* sometimes arise through the smoothing of an *irregular surface* (section 2.2a, and Discussion), which implies a structural influence, but this is not always the case. This is indicated by the fact that their relief is controlled by hydrodynamic conditions, being greater in areas of steeper and faster flow, such as immediately above steps and in short, steep reaches. These are also areas in which incision is greater generally, as the *hummocky forms* are surrounded by a high density of other sculpted forms. This implies a fundamental hydrodynamic control on the development of *hummocky forms* and further suggests that an undulating surface may develop spontaneously, even from an initially highly planar surface.

The influence of *hummocky forms* on flow patterns in shallow flow is interesting. While *nondirectional* forms set up a pattern of small hydraulic jumps on the stoss sides of the high points and, in the intervening areas, rhomboidal patterns of oblique shock

Figure 84. *Hummocky forms*. These examples have a slight longitudinal orientation, as shown by the depressions containing standing water, some of which are elongated parallel to flow. River Dee, UK. Flow from bottom right to top left. Limestone. The hammer is 30 cm long.

waves emanating from the hydraulic jumps, *transverse* forms set up wider hydraulic jumps trending across the channel and have fewer oblique shock waves. Such flow patterns are in fact present most of the time in headwater channels, which are flashy, and in which floods, which drown out these flow patterns, quickly subside. It is tempting to conclude that the hydraulic jumps are the agent responsible for the spatial variation in erosion rates that leads to the formation of *hummocky forms*. However, *hummocky forms* also form in pool-like areas of the channel in which hydraulic jumps are never present over the bedforms, and also on surfaces in the channel margin that slope steeply toward the channel center. Hydraulic jumps are therefore not essential to the formation of *hummocky forms* but will inevitably occur in shallow flows on *undulating surfaces* and probably act to enhance the formation of *hummocky forms*. For instance, Tinkler (1993) indicates the importance of hydraulic jumps in enhancing erosion locally. *Hummocky forms* may be classed as being of either low or high relief. An arbitrary threshold between these of 5 cm relief is suggested. It is thought that, as the relief of *hummocky forms* increases, the topographic highs of the bedforms begin to represent significant obstacles to the flow and develop attached vortices upstream of them, thus generating *obstacle marks* (section 3.1). This is suggested by the occurrence of *current crescents* within fields of high-relief *hummocky forms*. The terms "hummocky forms" and "hummocky topography" are sometimes used as a catchall phrase for complex sculpted bedrock surfaces of coalesced and interfering forms (e.g., Bryant, 2001: p. 89), but this is not the sense in which *hummocky forms* are defined here.

(a) *Nondirectional hummocky forms.* These show no orientation (i.e., the hummocks are equidimensional) or have both longitudinal and transverse forms present within a group. The "water-smoothed undulatory surfaces" of King (1927), the "undulating surfaces" (s-forms) of Kor et al. (1991) and Tinkler (1993), and the "undulose smoothing" (also an s-form) of Glasser and Nicholson (1998) may fall within this category.

(b) *Longitudinal hummocky forms.* These show an elongation in the streamwise direction, although it is never great (the length:width ratio of the hummocks does not exceed ~4) and the longitudinal nature of the forms is often subtle (Fig. 84). An area of *hummocky forms* must by definition contain depressions. It is therefore important not to confuse this bedform with, say, an area containing several *short longitudinal furrows*, especially if the rims of the *furrows* are rounded. An area of *longitudinal hummocky forms* is distinguished by its lack of closed depressions of a regular (i.e., elliptical) shape with well-defined rims. The depressions within an area of *longitudinal hummocky forms* are diffuse, ill-defined, and irregular in shape.

(c) *Transverse hummocky forms.* Transverse elongation. Rules similar to those described above apply for distinguishing *transverse hummocky forms* from an area containing several *transverse furrows*.

(d) *Sharp-crested hummocky forms* (*SCHF*). These are also elongated in a transverse direction, but unlike the three types so far described, which are rounded, *SCHF* have cuspate crests dividing gently sloping convex stoss slopes from steep, concave lee slopes (Figs. 85 and 127). The lee faces of *SCHF* are often darker than the stoss sides because of algal growth, indicating that abrasion there is less intense. In rocks with a pronounced structure, the structure interferes with the development of *SCHF*. In such cases, the crests may be parallel to structural trends rather than perpendicular to flow, and these would be termed *structurally influenced SCHF*. *SCHF* exist in a range of morphologies; three subdivisions are described here:

(i) *Restricted development of SCHF*: This is the poorly developed but common mode of the bedform. *SCHF* in this classification may occur as isolated individuals, or perhaps crests are found together but are irregularly spaced

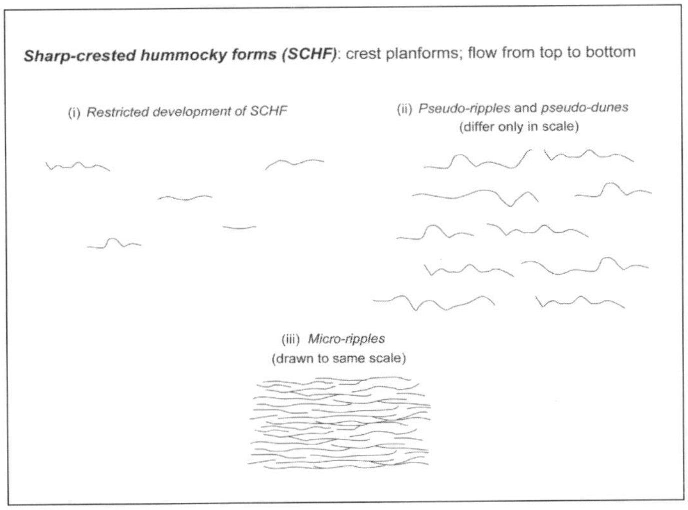

Figure 85. *Sharp-crested hummocky forms*.

and orientated or are few in number, or the crests may have low continuity (Fig. 86). The crests are typically sinuous and are of low relief. The origin of the crests is sometimes obvious. For instance, the type of *SCHF* with the lowest relief is that which develops in areas of *spindle-shaped flutes*, where the crests arise from the coalescence in a transverse direction of closely spaced *flutes*. A similar origin, but involving larger-scale *flutes*, is likely for some of the more typical, higher-relief *SCHF*, given that they are sometimes found to pass laterally into discrete *shallow flutes*, and that the *SCHF* in these cases contain crest sections that are parabolic in plan. In other cases, the origin is not apparent, and the crests simply grade into smooth bedrock. *SCHF of restricted development* are all but ubiquitous in bedrock channels on actively abrading surfaces where *furrows* are absent, wherever the substrate is relatively homogeneous and massive or where the reach has a *concordant plane bed*. This suggests that sharp-crested features (i.e., topographic singularities) are a stable state to which abrading bedrock surfaces tend (see Discussion).

(ii) *Regular train of SCHF* (*pseudo-ripples* and *pseudo-dunes*): This is the well-developed but more rare form of *SCHF*, although it is quite common in limestone. The bedforms have a higher relief and consist of many crests, apparently with some regularity to their spacing, giving the appearance of ripples or dunes depending on their scale (Fig. 87). They may therefore be called *pseudo-ripples* and *pseudo-dunes* respectively, for convenience. Previous descriptions of similar features include some of the "evorsion marks" of Ängeby (1951: e.g., Fig. 22 therein), the "hummocky surfaces" of Wohl (1992b), and the "ripple-like bedforms" of Hancock et al. (1998) and Whipple et al. (2000a). *SCHF* can also be seen in Figure 11 of Baker and Kale (1998), from the Katherine Gorge, Australia, although they are not described. Jennings (1983) describes a bedform that he calls "horns," which appear to be high-relief *SCHF* under the terminology of this report. Ashton and Kennedy (1972) described two-dimensional "ripples" on the underside of river ice covers. The ripples were initially rounded, but as their relief increased, they became more asymmetrical and developed sharp crests. However, the "transverse

Figure 86. *Restricted development of SCHF*. The transverse crests are short, isolated, and discontinuous, and do not form part of a train of bedforms. Flow from left to right. Than Rattana, Thailand. Andesite. The notebook is 15 cm long.

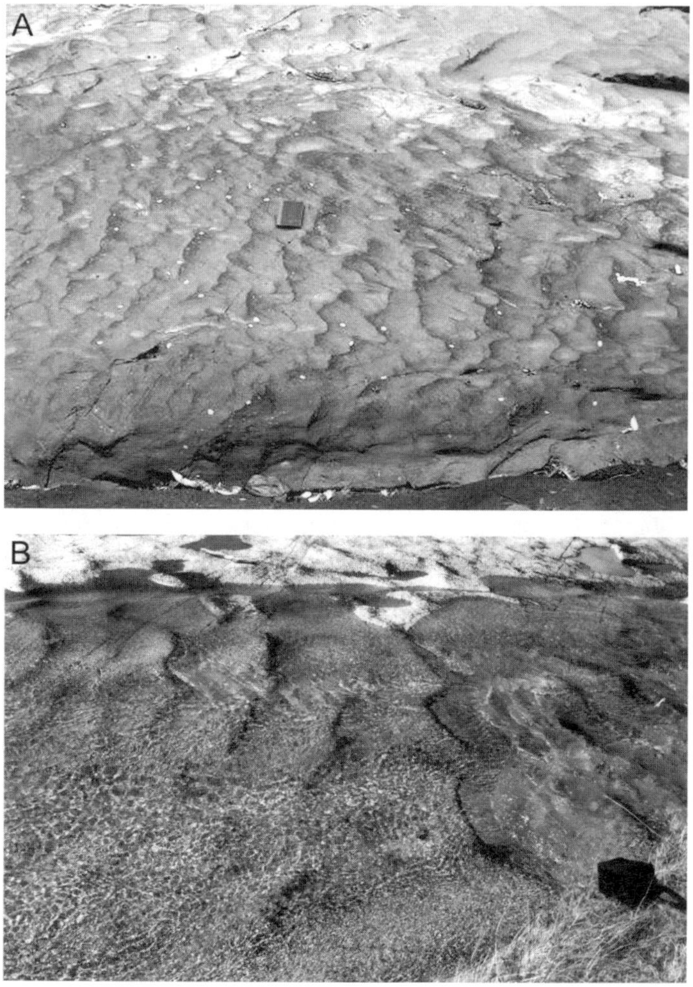

Figure 87. A: *Pseudoripples*. Than Rattana, Thailand. Flow from left to right. Andesite. The notebook is 15 cm long. B: *Pseudodunes*. Allt Ceitlein, UK. Flow from left to right. Granite. Camera bag, 20 cm across, for scale.

erosional ripples" of Shepherd and Schumm (1974: Fig. 4 therein) appear in fact to be *transverse furrows* within an otherwise planar surface.

Pseudoripples and *pseudo-dunes* are variable in character, and under this umbrella term it is possible that a range of different bedforms is included. Like *directional scallops* (section 4.2b), they vary from highly three-dimensional to relatively two-dimensional forms. Variation in the third dimension may simply take the form of irregular perturbations and discontinuity in the crests, or it may be more regular and take the form of repeating arcuate crest sections that in plan view are convex-upstream and that contain hollows similar to *flutes* (Fig. 88). In such cases, they are similar in appearance to *three-dimensional directional scallops* (section 4.2b), and it is possible that they evolve along a similar path, i.e., via a conjugate assemblage of *flutes* (see Discussion). However, *SCHF* are a diverse group of forms that are in general completely distinct from *scallops*, and that show continuity and an inferred relationship with rounded *hummocky forms*. Furthermore, Hancock et al. (1998) observed that when *pseudo-dunes* attain sufficient relief, their crests spawn *flutes* that attack the lee faces, suggesting that in some cases *flute*-like hollows within *SCHF* are secondary features. *Pseudoripples* that occur with well-defined and recognizable *short furrows* in their *troughs* are treated as a *composite form* and discussed separately (section 3.2a). It has been noted by the authors that an increase in gradient and flow velocity (and therefore a reduction in depth) in the channel causes the following modifications in the morphology of *pseudo-dunes*:

- Relief increases.
- The stoss sides become steeper.
- The *SCHF* become more three-dimensional and discontinuous, and they tend to become arranged in discrete patches that form mounds, separated by *parallel sided furrows* (see section 3.2a).
- As a result of the increase in relief and the steepening of the upstream faces, some *SCHF* begin to behave as significant obstacles to the flow, developing associated *current crescents* (section 3.1), and some *troughs* become deepened and take on the characteristics of *simple transverse furrows*, with the *SCHF* themselves acting as the "defects" responsible (sections 1.3a and 3.2c).

(iii) *Microripples*: On some highly abraded rock surfaces, very small-scale ripple-like forms can sometimes be observed (Fig. 89). These tend to occur in areas of high gradient and/or converging streamlines, such as in rapids, on *chutes*, and immediately above *steps*. These occur on a range of scales, from a wavelength of ~1 cm up to the size of typical *SCHF*, with which they appear to form a continuum. They often occur in a hierarchical arrangement, with *microripples* being superimposed on typical *SCHF* with a wavelength of decimeters. In such cases, the *microripples* form preferentially on the stoss faces of the *SCHF*. This is in contrast to the development of *solution pitting* on *hummocky forms* in limestone, which occurs preferentially on the lee faces. In planform, the crests of the *microripples* are discontinuous and are generally sinuous in an irregular fashion, although in some cases they show repeating arcuate and angular crest sections. The arcuate sections are convex in the upstream direction, and the angular sections point downstream, and these cases are, again, similar in appearance to *directional scallops*.

2.2 Other Convex and Undulating Surfaces

(a) *Partially abraded surface.* This term applies to surfaces in which a combination of plucking and/or subaerial weathering and fluvial abrasion and/or dissolution have been active. Plucking and subaerial weathering produce angular features and *irregular*

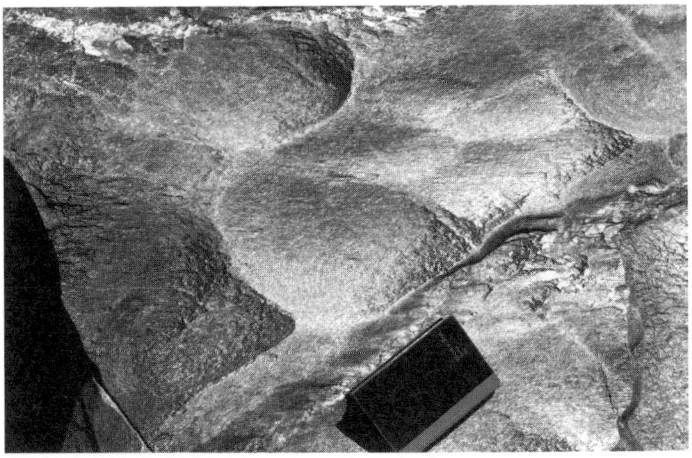

Figure 88. *Pseudoripples* of a three-dimensional kind. Note the presence of parabolic segments in the crests, enclosing *flute*-like features. Flow from right to left. Nam Mae Khlang, Thailand. Gneiss. The notebook is 15 cm long.

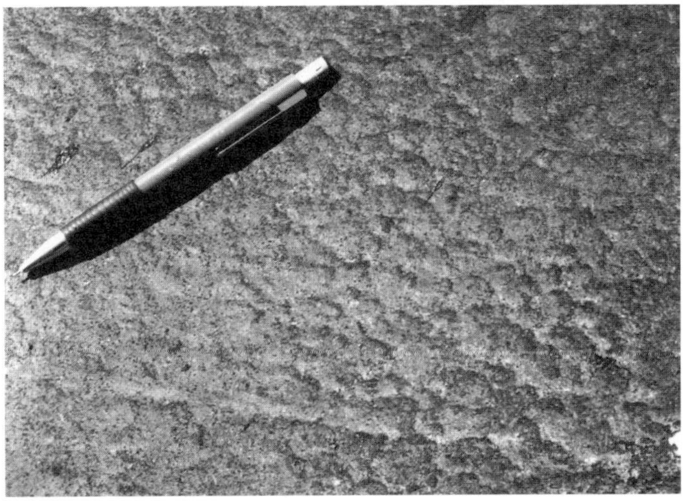

Figure 89. *Microripples*. Flow from top right to bottom left. Nam Mae Khlang, Thailand. Gneiss. Pen for scale.

surfaces that subsequently become subject to fluvial abrasion or dissolution, and unless the rate of removal of material by plucking/weathering is especially rapid, these surfaces will be partially smoothed and rounded, with the abrasion or dissolution being most intense on the upstream side of high points. The resulting *partially abraded surfaces* are variable in appearance. They range from surfaces consisting of subangular corners and projections formed by the intersection of joint and bedding planes, to *irregular surfaces* in which the upstream sides of high points have been worn smooth, leaving the lee sides rough (Fig. 90). Incipient and small sculpted forms may also be present. As one proceeds toward the thalweg or down the channel margins, a gradual increase is often found in the amount of fluvial abrasion/dissolution that has occurred relative to plucking and subaerial weathering, manifested by a gradual increase in the proportion of an *irregular surface* that has been smoothed, or an increase in the degree of rounding of angular corners and projections. Sculpted forms also become better developed. Complete removal of an *irregular surface* by abrasion generally results in an area of *hummocky forms*, which may be either rounded or *sharp-crested*. It is thought that if a high point in a *partially abraded surface* is at any time sufficiently steep to cause flow separation on the lee side, it develops into a sharp-crested hummocky form; otherwise it will be a rounded hummocky form. In chutes and in very steep channels (slope >10°, say), areas of alternating rough and smooth rock with sharp boundaries may result from the complete separation of the water from (i.e., ventilation of) the rock surface at irregularities and changes in surface orientation. Ventilation prevents abrasion immediately downstream of the point of separation from the rock surface, and this point forms the boundary between rough and smooth rock.

(b) *Polished surface.* This term describes the extremely smooth and shiny, gently *undulating surfaces* (Fig. 91) formed by abrasion by suspended sediment on hard, dense rocks that are not too coarse-grained (e.g., medium sandstone and finer). The appearance may be similar to that of burnished copper. In conglomerates, individual clasts may be polished, the degree of which varies with lithology. *Polished surfaces* are similar to the patinas that develop subaerially on rocks in arid conditions owing to wind-induced abrasion. *Polished surfaces* appear to be hardened external layers that resist erosion and occasionally are broken through or flake off, exposing rough, softer rock beneath, which is more rapidly eroded. *Polished surfaces* may cover small outcrops and rock highs, and they are sometimes developed

Figure 91. *Polished surface* developed on a bedrock high. Note the sharp delimitation of polished and rough surfaces (arrowed). Som Ploy River, Thailand. Flow from top right to bottom left. Fine-grained sandstone. Coin in center of photo is 20 mm across.

Figure 90. *Partially abraded surface*. Gayle Beck, UK. Flow from top to bottom. Limestone. The scale is 60 cm long.

Figure 92. *Upstream-facing convex surfaces*. River Wharfe, UK. Flow from right to left. Limestone. The scale is 60 cm long.

within *concave features*. *Polished surfaces* are described by Baker and Pickup (1987), Baker and Kale (1998), Hancock et al. (1998), Gupta et al. (1999), Spaggiari et al. (1999), and Whipple et al. (2000a).

(c) *Upstream-facing convex surface.* This is the general form of the upstream faces of boulders and rock projections and consists of a very smooth and gently rounded surface (Fig. 92), often with an abrupt and angular downstream edge (see section 3.3). *Upstream-facing convex surfaces* were featured prominently by Hancock et al. (1998: e.g., Fig. 5 therein) and Whipple et al. (2000a). They represent sculpting within the zone of convergent streamlines on the upstream face of an obstacle to the flow. Surfaces thus sculpted lose any angularity and become streamlined.

(d) *Irregular sculpting.* The steeply dipping lee faces of boulders and rock projections are commonly found to possess small-scale (centimeters to decimeters) irregular, rounded undulations of low relief, especially in limestone or calcareous argillaceous rocks (Fig. 93). This is clearly the effect of sculpting within a region of recirculating flow. As steeply dipping faces of slabs and projections become orientated more parallel to the flow, the sculpting becomes more organized and becomes transformed into *short longitudinal furrows* or *SCHF*. It may also be more organized if the rock has an internal structure. *Irregular sculpting* may also be found on upstream-facing surfaces if these are in the wake of some other obstacle and, in channel margins with a square cross section, on horizontal surfaces close to the right-angled corner of the cross section.

(e) *Convoluted surface.* This is something of a catchall for complex, smoothly curved surfaces that result from the intense sculpting of bedrock, producing various interfering and coalescing *furrows* and *convex surfaces* that can no longer be discerned.

(f) *Bladed forms.* These are high-relief, sharp-crested, generally longitudinal ridges formed by and between two contiguous *furrows* (Fig. 94).

(g) *Faceted obstacle.* The faceting of obstacles to the flow involves the erosion of a boulder, pebble, or rock projection to produce approximately planar surfaces. Upstream-facing facets analogous to those familiar from aeolian environments were described by Maxson (1940) from the Colorado River and by Baker and Pickup (1987) from the Katherine Gorge, Australia. However, in contrast to aeolian faceting, flow-parallel facets are also developed in streams. These intersect in a sharp longitudinal keel (Fig. 95). Bryant (2001: p. 87) described *faceted obstacles* (which he referred to as "flutes") from coastal environments due to tsunamis.

3. Composite Forms

This is the third major class of sculpted forms, and it comprises features that consist of *concave features* and *convex* or *undulating surfaces* combined within a single bedform, i.e., the bedforms contain components from both of the previous two classes. "Composite" is used here in a subtly different sense to the term "compound," described in the terminology section. Compound forms consist of *concave features* subdivided internally into further, smaller *concave features*. *Composite forms*, on the other hand, are composed of two or more morphological components of an entirely dissimilar nature. The development of

Figure 93. *Irregular sculpting* (arrowed) on the lee face of a rock projection. River Lune (Tebay), UK, looking upstream. Calcareous mudstone. The hammer is 30 cm long.

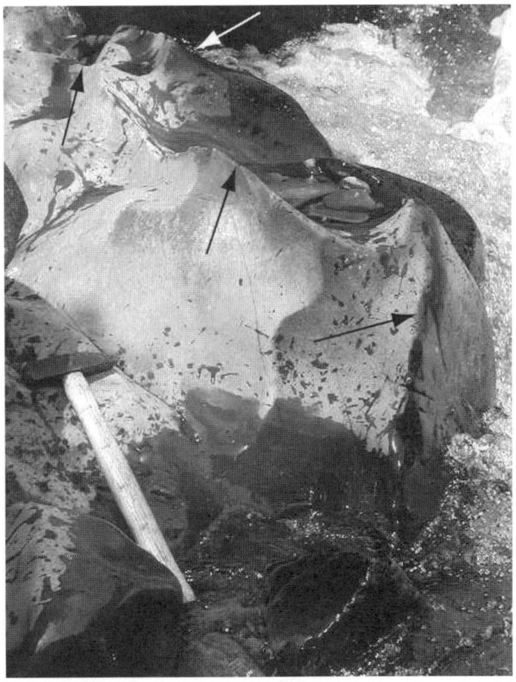

Figure 94. Several *bladed forms* (arrowed) on a rock projection. River Lune (Tebay), UK, looking upstream. Calcareous mudstone. The hammer is 30 cm long.

Figure 95. *Faceted obstacle* in a channel margin, seen from above. Three planar faces (one of which the hammer is resting upon) are present, and these produce a pointed top to the obstacle. River Lune (Tebay), UK. Upstream to left. Calcareous mudstone. The hammer is 30 cm long.

these components is, and probably always has been, intimately related, and they are indivisible morphologically and functionally. The components often behave in a mutually constructive manner, and together they form a new bedform.

3.1 Obstacle Marks

Obstacle marks are well known from both cohesive and noncohesive sedimentary deposits (Peabody, 1947; Dzulynski and Sanders, 1962; Dzulynski and Walton, 1965; Karcz, 1968, 1973; Allen, 1974, 1982: p. 177–196) and snow (Allen, 1965) and have also been described in bedrock (Baker, 1973; Lorenc et al., 1994). They result from secondary flow structures induced by the presence of a significant obstacle to the flow, which create a crescentic *furrow* (*current crescent*) around the *obstacle*. Sedimentary examples may involve both scour and deposition (Allen, 1982: p. 184). Although *current crescents* in bedrock sometimes occur around engineering structures and nontransported boulders, in general the *obstacle* is simply a projecting part of the outcrop and is an integral part of the *obstacle mark*. This contrasts with sedimentary *obstacle marks* where the *obstacle* is a "foreign" body such as a pebble in sand. This is possible because the rock has strength and can form its own *obstacles*, whereas sediment cannot. In bedrock, the *obstacle* sometimes consists of a bedform, such as a *hummocky form*, which has attained high relief. The "streamside potholes" of Ängeby (1951) are sculpted forms occurring on the upstream side of an *obstacle* and would be included with *obstacle marks* in this typology; similarly, some of the "transverse grooves" of Jennings (1983) are also *obstacle marks*.

Obstacle marks in bedrock channels are defined by two elements: a smoothed and rounded *obstacle*, and an associated

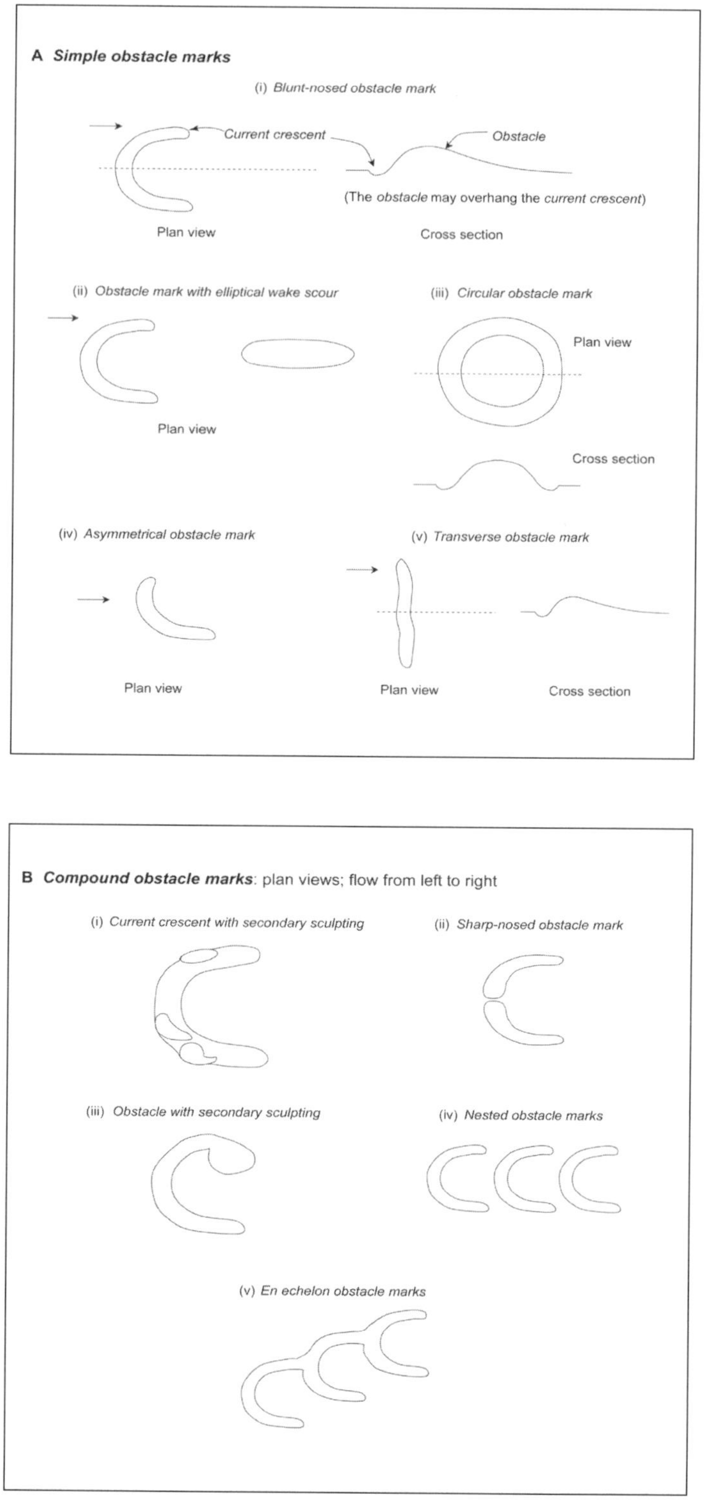

Figure 96. *Obstacle marks*. Arrows indicate flow directions (where appropriate). Dashed lines indicate lines of cross sections.

current crescent (Peabody, 1947) situated on the upstream side and passing down either side of the *obstacle* (Fig. 96). The *obstacle* is analogous to an *upstream-facing convex surface*, and the *current crescent* is a crescentic *transverse furrow*. In mature, well-developed examples, the sides of the *current crescent* in the upstream part may overhang. At its downstream end, the *obstacle* typically terminates sharply at a joint face or where other sculpted forms are eroded into it, but in ideal cases the *obstacle* is elongated and tapers in the downstream direction while losing height. Such features are analogous to the *current shadows* formed by differential erosion in the lee of *obstacles* in scour marks in cohesive sediments (Allen, 1982: Fig. 5-10 a, f, g therein), although in bedrock they are integral with the *obstacle*. *Current shadows* are more distinct in *obstacle marks* formed in bedrock around engineering structures (Fig. 97). The *current shadow* reduces the amount of flow separation and represents a form of streamlining of the lee side of the *obstacle* (Richardson, 1968). The *current crescent* is sculpted by a standing horizontal horseshoe vortex whose arms embrace the *obstacle* (Richardson, 1968; Karcz, 1968, 1973; Allen, 1982: p. 177–178; Acarlar and Smith, 1987). Evidence for this can be seen in small-scale sculpting within the *current crescents* of some *obstacle marks*, which indicates the skin friction lines and therefore the flow structures present when the bedform is active (Allen, 1965) (see also Fig. 167). A range of morphologies of *obstacle marks* in bedrock can be identified.

(a) *Simple obstacle marks.* This group consists of isolate marks that are unmodified by secondary sculpting.

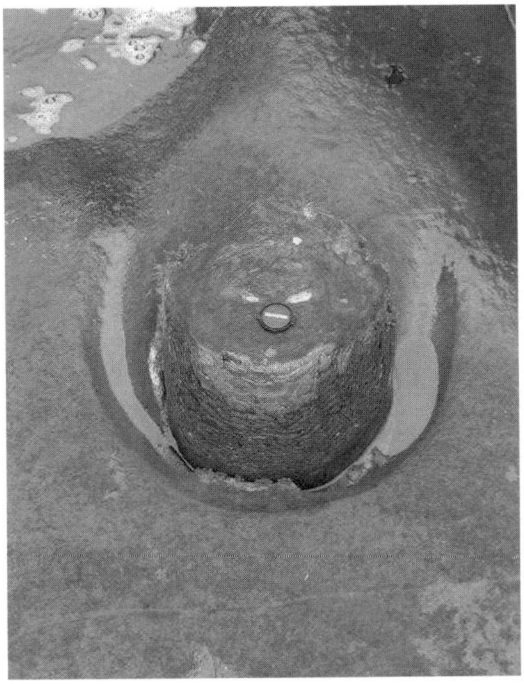

Figure 97. *Current crescent* around concrete post set in bedrock. Note the well-developed current shadow (top). Sichiri-gawa, Japan. Flow from bottom to top. Fine-grained sandstone. Lens cap for scale.

(i) *Blunt-nosed obstacle mark*: This is the basic and most common form of *obstacle mark*. The *obstacle* is gently rounded at its upstream end (Fig. 98).

(ii) *Obstacle mark with elliptical wake scour*: This variety was described by Baker (1973). It consists of a large elliptical scour hole in the wake of the *obstacle* (in this case a boulder), in addition to the *current crescent* upstream of it. Lorenc et al. (1994) also describe *obstacle marks* with scour in the wake of boulders.

Figure 98. *Blunt-nosed obstacle mark*. In this example, the *obstacle* is slightly overhanging on its upstream side (arrowed). Nam Mae Chaem (Ob Luang), Thailand. Flow from right to left. Granitic gneiss. The notebook is 15 cm long and is resting on a boulder situated within the *current crescent*.

Figure 99. *Circular obstacle mark* (to right of notebook). The *obstacle* in this case is a coarse clast within a rhyolitic agglomerate. Huai Nang Rong, Thailand. Upstream to right. The notebook is 15 cm long.

(iii) *Circular obstacle mark*: Sometimes the *current crescent* forms a complete ring around the *obstacle* (Fig. 99), which is radially symmetrical, indicating the presence of rotary or multidirectional currents (Allen, 1982: Fig. 5-9). In bedrock channels, this probably represents an area of extreme macroturbulence.

(iv) *Asymmetrical obstacle mark*: This consists of a *current crescent* that is more developed on one side than the other (Fig. 100). It may be analogous to the comma form (type of s-form) of Kor et al. (1991) and Shaw (1996).

Figure 100. *Asymmetrical obstacle mark*. Nam Mae Chaem (Ob Luang), Thailand. Flow from bottom to top. Granitic gneiss. The notebook is 15 cm long.

(v) *Transverse obstacle mark*: Occasionally, flow separation upstream of an *obstacle* with considerable transverse extent results in an *upstream-facing convex surface* with a relatively straight *transverse furrow*, rather than a *current crescent*, on its upstream side (Fig. 101). Many of the "streamside potholes" of Ängeby (1951) are bedforms of this type.

(b) *Compound obstacle marks*. This group includes more complex *obstacle marks* containing secondary sculpting. It also includes swarms of closely packed and interfering *obstacle marks* that may be found in areas of intense erosion, two arrangements of which can be identified.

(i) *Current crescent with secondary sculpting*: Some well-developed (i.e., mature and of high relief) *current crescents* contain one or more smaller-scale secondary sculpted forms within them, such as *potholes* and *short furrows* (Fig. 102). This secondary sculpting is initially restricted to the upstream portion of the *current crescent*, however, and the *obstacle* is largely unmodified, remaining blunt-nosed and rounded. However, the secondary sculpted form(s) can grow to dominate the bedform, such that the initial *obstacle mark* is difficult to identify. Figure 15, for example, shows a well-developed *compound pothole with extended exit furrow* sited in a *current crescent*, the latter of which is no longer obvious and may now be moribund. A *current crescent with secondary sculpting* is illustrated by Nemec et al. (1982: Plate I, Photo 4 therein),

Figure 101. *Transverse obstacle mark*. Nam Mae Chaem (Ob Luang), Thailand. Flow from right to left. Granitic gneiss. The notebook is 15 cm long.

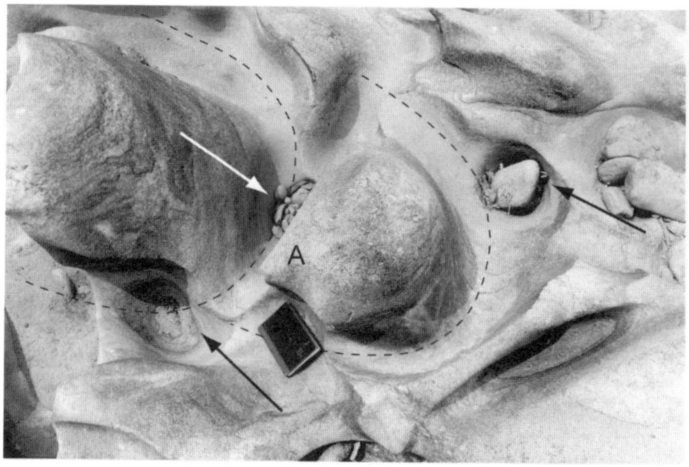

Figure 102. *Current crescents with secondary sculpting*. Two *obstacle marks* are shown, and the associated *current crescents* (dashed lines) contain secondary sculpting (arrowed) in the form of *potholes* and a *short furrow*. The *obstacle marks* are *en echelon*, the downstream one being displaced to the right and eroding into its upstream neighbor, creating an angular downstream margin (A) to it. Nam Mae Chaem (Ob Luang), Thailand. Flow from top right to bottom left. Granitic gneiss. The notebook is 15 cm long.

who described *current crescents* as being common sites for the inception of *potholes*.

(ii) *Sharp-nosed obstacle mark*: Sometimes, secondary sculpting within the *current crescent* takes the form of two equal-sized *potholes* or *deep short furrows* positioned symmetrically about the dividing plane of the *current crescent* (Fig. 103). This secondary sculpting has the effect of producing a "sharp nose" to the *obstacle*, i.e., a longitudinal ridge at its upstream end formed by the crest dividing the two secondary sculpted forms.

(iii) *Obstacle with secondary sculpting*: In this case, the secondary sculpting occurs in the flank(s) of the *obstacle*. It is inferred that at one or more points, the radius of curvature of the *current crescent* (in the horizontal plane) is sufficiently small that flow separation occurs about a vertical axis along the side of the *obstacle*. This has the effect of increasing the rate of erosion in the area of reattachment and downstream of it, of shifting the locus of the maximum rate of erosion from the floor of the *current crescent* to the side of the *obstacle*, and thus reorientating the arm of the *crescent* in which flow separation occurs. It also modifies the shape of the *obstacle* by eroding into it from the side and producing in the flank of the *obstacle* a secondary sculpted form that is essentially a small *lateral pothole*. The *lateral pothole* has a sharp, vertical upstream rim that represents the point of flow separation (Fig. 104A).

(iv) *Nested obstacle marks*: *Obstacle marks* are frequently found in nested formations of interfering forms, in which each *current crescent* is eroded into the distal part of the next *obstacle* upstream, forming a longitudinal, linear sequence (Figs. 103 and 104B). The *current crescents* remain isolate, but closely spaced longitudinally. *Nested obstacle marks* can be seen in the illustrations of Jennings (1983: Plate 5 therein).

(v) *En echelon obstacle marks*: In this case, the marks are again in a linear formation, but each is translated in a consistent stepwise fashion with respect to its upstream neighbor, forming a sequence that is oblique to the flow direction. One flank of each *obstacle* in the sequence has been eroded into, but in this case by the *current crescent* of the next *obstacle mark* downstream (Fig. 102), rather than by a *lateral pothole* as was the case in an *obstacle with secondary sculpting*. Both *nested* and *en echelon* sequences

Figure 103. *Sharp-nosed obstacle mark*. The *obstacle*'s sharp nose (arrowed) divides the *current crescent* into two discrete halves. A *blunt-nosed obstacle mark* can be seen immediately to the left of the notebook, and these two *obstacle marks* exhibit a *nested* arrangement. Nam Mae Chaem (Ob Luang), Thailand. Flow from right to left. Granitic gneiss. The notebook is 15 cm long.

Figure 104. A: *Obstacle with secondary sculpting*. The inferred original course of the *current crescent* is along the black line. The present (modified) course has caused it to erode into the *obstacle*, resulting in a sharp edge (S) in the *obstacle*'s left side. Flow separation occurs at the sharp edge and has produced an incipient *pothole* there. There are cobbles in the *current crescent* next to the notebook. Nam Mae Chaem (Ob Luang), Thailand. Flow from right to left. Granitic gneiss. The notebook is 15 cm long. B: *Nested obstacle marks*. Two *current crescents* can be seen, one on either side of the pen, in a *nested* arrangement. Nam Mae Chaem (Ob Noi), Thailand. Flow from left to right. Granitic gneiss.

suggest that primary marks are able to generate external secondary marks as a result of the flow structures they produce, in a manner similar to that described above for *flutes with external secondary structures*.

3.2 Hummocky Forms with Linear Depressions

These bedforms are an intimate association of *hummocky forms* and *furrows* or *runnels*, indicative of their closely related development.

(a) *Pseudoripples with longitudinal furrows.* It is common for *pseudo-ripples* and *pseudo-dunes* to occur in association with *longitudinal furrows*. This association appears to be promoted by a high gradient and faster flow, and may occur in either, or both, of two ways:

(i) *Pseudoripples with short furrows*: The *furrows* occur in the *troughs* of the *pseudo-ripples*, generating high crest sinuosity as the *furrows* deflect the crests around themselves (Blank, 1958: Fig. 3 therein; Sevon, 1988: Figs. 2 and 4 therein; Whipple et al., 2000a: Fig. 2A therein) (Fig. 105A). The crests of the *pseudo-ripples* are either partly or completely formed from, and consist of, the crests between contiguous *short furrows*. This indicates not merely a close relationship between *SCHF* (*sharp-crested hummocky forms*) and *longitudinal furrows*, but an identity in this case. Where the crests of the *pseudo-ripples* are formed completely in this way, the bedform may be viewed either as *pseudo-ripples with short furrows* or as a field of conjugate *short furrows*; the two are equivalent (Fig. 105B). The presence of a number of *short furrows* in *troughs* has the effect of introducing some crest sections parallel to the flow direction (i.e., where two such *furrows* are contiguous side by side), in addition to the dominant sinuous transverse crests of the *pseudo-ripples*. However, the (average) orientation of the *pseudo-ripple* crests and that of the axes of the *short furrows* are sometimes interfered with by the rock structure and are not always perpendicular. A number of contiguous, in-line *short furrows* might be considered to form a *compound parallel-sided furrow* (section 1.2d).

(ii) *Pseudoripples with parallel-sided furrows*: In this case, the *furrows*, which are longer than the wavelength of *pseudo-ripples*, tend to cut across crests (Fig. 106A). This bedform may represent a coexistence of the two types of features in an environment that is transitional between that which would produce only *pseudo-ripples*, and that in which only *parallel-sided furrows* would be developed. This would make the bedform directly analogous to the coexistence of ripples and harrow marks described by Karcz (1967) in sand. In some cases, the *furrows* define larger-scale, second-order structures (mounds) on which the *pseudo-ripples* occur (Fig. 106B).

(b) *Runnels with bed relief.* *Runnels with wall relief* have already been described (section 1.2 f). *Runnels with bed relief* consist of *runnels* whose floors contain quasi-periodic topography of varying relief, with varying effects on the flow pattern within the *runnel*. The following examples are arranged in order of the impact of the bed relief on the flow in the *runnel*.

(i) *Runnel with hummocky forms*: This topography has the lowest relief and has little effect on the flow; it is not sufficient, for example, to create hydraulic jumps or standing waves. The *runnel* bed consists of smooth and gently undulating *hummocky forms*.

(ii) *Runnel with SCHF*: The topography in this case is of higher relief, and the crests of the *hummocky forms* within the *runnel* are sharp. The crests may also be laterally continuous with *SCHF* on either side of the *runnel*. Where the *runnel*

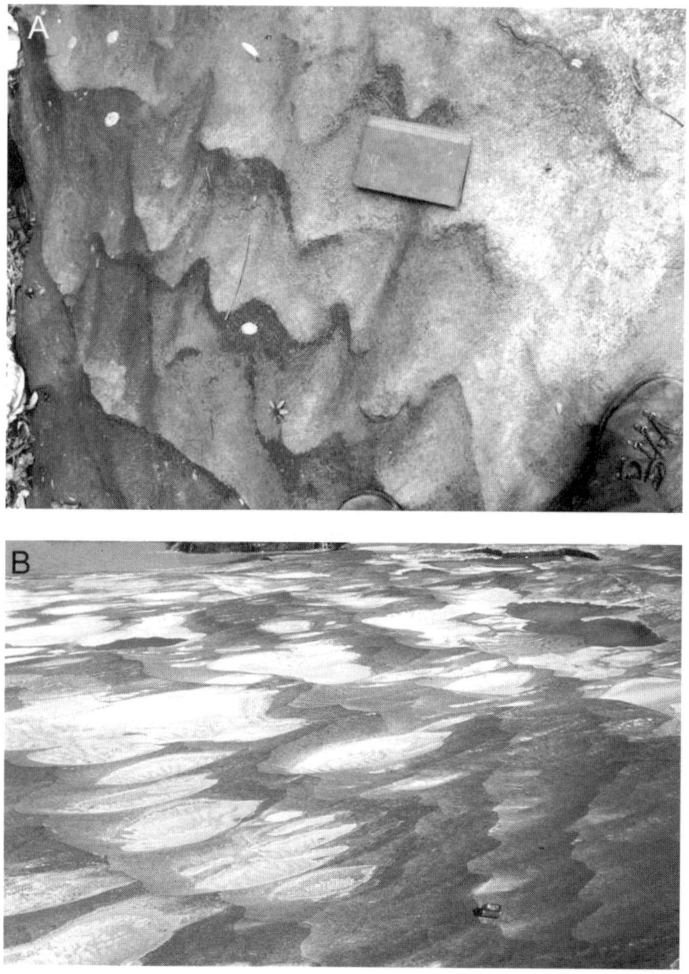

Figure 105. A: *Pseudoripples with short furrows*. The *short furrows* exist in the *troughs*, and the *pseudo-ripple* crests are deflected around them, becoming sinuous. The sinuous transverse crests are in fact formed in part as the crests between conjugate *short furrows*. Than Rattana, Thailand. Flow from top to bottom. Andesite. The notebook is 15 cm long. B: *Pseudodunes with short furrows*. Here, the transverse crests are formed almost entirely through the conjugation of *short furrows*. The pattern may therefore be viewed equivalently either as three-dimensional *pseudo-dunes* or as an assemblage of conjugate *short furrows*. There are some flow-parallel crest segments, and the transverse crests are less dominant than in Figure 105A. Watson River, Greenland. Flow from top right to bottom left. Gneiss. Tape measure in foreground for scale. Photo courtesy of Andy Russell.

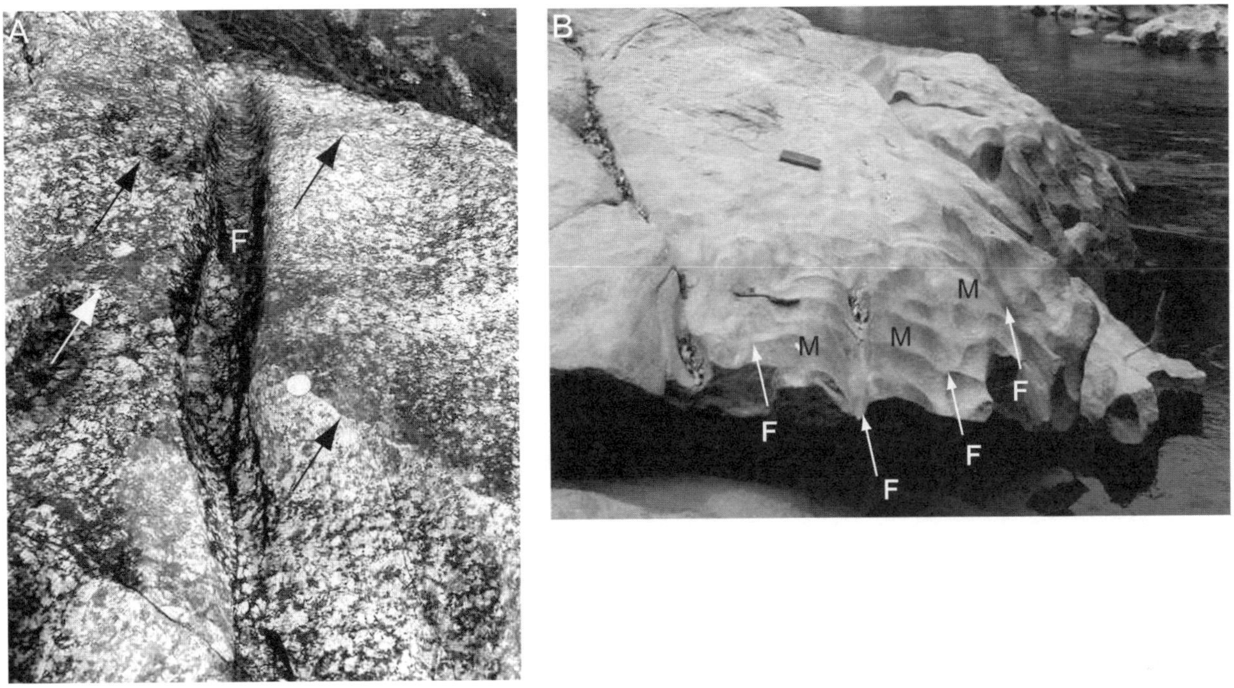

Figure 106. A: *Pseudoripples with parallel-sided furrow*. Note the presence of the *furrow* (F) cutting across *pseudo-ripple* crest lines (arrowed). Woolshed Creek, Australia. Flow from bottom to top. Granite. Coin for scale. B: *Pseudoripples with parallel-sided furrows*. In this example, the *parallel-sided furrows* (F) define intervening mounds (M) on which the *pseudo-ripples* occur. Nam Mae Chaem (Ob Luang), Thailand. Flow from top to bottom. Granitic gneiss. The notebook is 15 cm long.

changes orientation within the main channel, the *SCHF* within the *runnel* change orientation with it, remaining approximately transverse to it and indicating that they are formed by low-stage flow channeled within the *runnel* (see section 1.2f). In plan view, the *SCHF* may be sinuous, they may cross the *runnel* obliquely, or they may be V-shaped (with the V pointing downstream); crests that are straight and purely transverse are rare (Fig. 107). *Parallel-sided furrows* and *short furrows* may also be present within the *troughs* of the *SCHF*, either parallel to the *runnel* or parallel to the main channel (if different). Hydraulic jumps and standing waves are present over the crests within the *runnel* at low to moderate stages. This bedform differs morphologically from a *regular compound parallel-sided furrow* (section 1.2d) in that the latter is deeper relative to its width, and consists of a longitudinal series of well-defined, contiguous, component *parallel-sided furrows*. A *runnel with SCHF* has more poorly defined *troughs* between the sharp crests, which have a width comparable to or greater than their wavelength, and which therefore do not resemble a series of *parallel-sided furrows*. These bedforms also differ in that a *runnel with SCHF* acts as a conduit for channeling low stage flow when surrounding bedrock is very shallow or even dry.

Figure 107. *Runnel with SCHF*. The sharp crests within the *runnel* vary from purely transverse (T) to oblique (O) and from straight (St) to sinuous (Si). Nam Mae Khlang, Thailand. Flow from top to bottom. Gneiss. The notebook is 15 cm long.

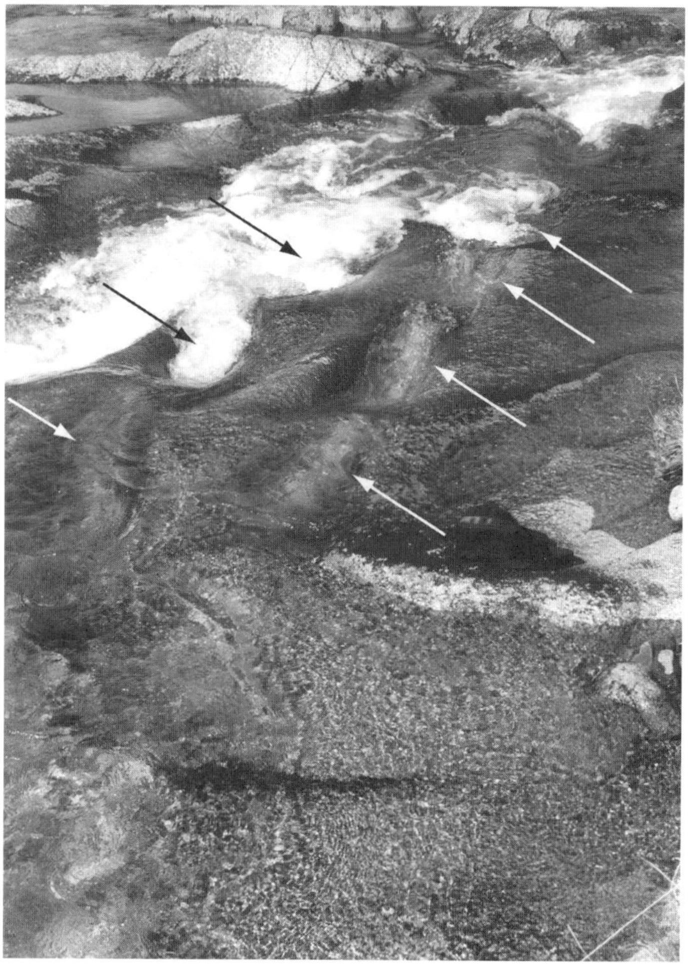

Figure 108. Parallel *runnels with step-pool structures*. The pools are arrowed. Allt Ceitlein, UK, looking downstream. Granite. Camera bag, 20 cm wide.

(iii) *Runnel with step-pool structures*: When the topography within the *runnel* is of even higher relief and/or the *runnel* is of higher gradient, the flow within the *runnel* is best described as having a *step-pool* structure (Fig. 108; see also section 6.2g), and the typical associated flow pattern consists of a pronounced alternation of supercritical and subcritical flow, with hydraulic jumps. This bedform is probably also formed by low-stage flow channeled within the *runnel* because at moderate to high stages the *step-pools* would be drowned out. Some of the *furrows* described by Blank (1958: e.g., Fig. 4 therein) from the James River, Texas, are of this type. Wohl and Ikeda (1997) also generated *step-pool* structures in high-gradient (40%) runs in their flume experiments. The topography and flow structure of a *runnel with step-pool structures* are similar to those of the sedimentary bedform known as a chute-and-pool structure (Leeder, 1982: p. 92; Fralick, 1999). The terms chute-and-pool and *step-pool* are considered synonymous in the context of bedrock channels in this typology.

(c) *High-relief hummocky forms with transverse furrows.* It has already been noted (section 2.1) that the relief of *hummocky forms* increases in reaches of higher gradient and above *steps*. If they achieve sufficient relief, the *hummocky forms* can strongly modify the flow and erosion patterns around them.

(i) *High-relief SCHF with transverse furrows*: Here, the *SCHF* act as the "defects" necessary for the generation of a roller and *transverse furrow* on their lee sides (Fig. 109).

(ii) *High-relief hummocky forms with current crescents*: In this case, the *hummocky forms* (of any type) begin to act as a significant obstacle to the flow, generating flow separation and an associated *current crescent* on their upstream side (Fig. 110).

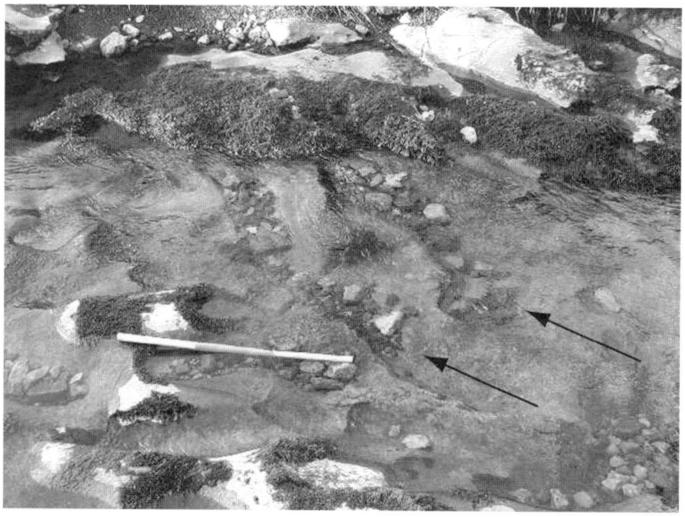

Figure 109. High relief *SCHF with transverse furrows*. The *furrows* (arrowed) are partially filled with pebbles. River Wharfe, UK. Flow from left to right. Limestone. The scale is 60 cm long.

Figure 110. High relief *Hummocky forms with current crescents*. Two *hummocky forms* (one at either end of the pole, and partly covered with moss) have begun to behave as *obstacles* and have generated *current crescents*. River Wharfe, UK. Flow from right to left. Limestone. The scale is 60 cm long.

3.3 Convex Surfaces with Steep Lee Faces

It is common for sculpted *convex surfaces* to terminate at, or contain within them, *steep* (nearly vertical to slightly overhanging) *lee faces*. This can occur on a range of scales, from large *upstream-facing convex surfaces* developed on boulders and projections, to the relatively small convex areas of *hummocky forms*. The *steep lee face* normally arises through (1) the extensive and rapid erosion of a surface by *flutes*, *potholes*, etc., just downstream of a convex rise, and (2) their coalescence to form a near-vertical lee face to that rise. Hancock et al. (1998) and Whipple et al. (2000a) describe this process on boulders and projections, which tend to have smooth, rounded upstream sides, *fluted* flow-parallel sides, and heavily *fluted* and *potholed* lee faces. The near-vertical lee faces are convoluted with cuspate projections, indicating their origin through breached and coalesced *flutes* and *potholes*. Once developed, *steep lee faces* may become host to secondary *irregular sculpting*. Two specific examples of the formation of *convex surfaces with steep lee faces* have been observed by the authors.

(a) *Hummocky form with steep lee face.* These have been observed to occur through two related mechanisms:

(i) Coalescence of *solution pits*. On a small scale, *steep lee faces* can occur on high-relief *hummocky forms* in limestone, where *solution pits* (section 4.1a) are developed preferentially on the lee faces. Pitting can become quite intense and leads to the coalescence of the *solution pits*, deep incision, and the development of *steep lee faces* on the scale of the original *hummocky form*. In well-developed examples, there are no relict crests (between pits) or small depressions as evidence of their origin from *solution pits* (Fig. 111A).

(ii) Coalescence of *potholes*. At Gayle Beck (Table 2), the bed consists of *hummocky forms*, on which *solution pits* develop on the lee faces (Fig. 159). With continued growth, the *solution pits* may become small *potholes*, the coalescence of which, in a transverse direction, results in a low *convex surface with steep lee face* of a transverse extent considerably greater than the original *hummocky form* (Fig. 111B). This illustrates how near-vertical *backward-facing steps* can arise even in a relatively flat bed devoid of prominent irregularities, and how such beds can consequently become considerably rougher. It is possible that within this category could be included the "helms" of Jennings (1983), which appear to be high-relief *hummocky forms* with *potholes* on the downstream side.

(b) *Streamlined steep lee face.* This *composite form* consists of an *upstream-facing convex surface*, the lee face of which is sculpted by two *concave features* (typically *lateral potholes* or *oblique sloping furrows*) into a streamlined form. In plan view, the *convex surface* has an angular section at its downstream end, formed by the two *concave features* (Fig. 112). At the apex of the angular section, where the two *concave features* intersect, is a sharp, near-vertical crest. The angular, downstream-pointing form of the lee side, ending in a sharp edge analogous to a tail fin,

Figure 111. A: *Hummocky forms with steep lee faces* developing through the coalescence of *solution pits* on the downstream sides of high-relief *hummocky forms*. The origin of the *steep lee faces* (arrowed) is indicated by their pitted nature. River Dee, UK. Flow from top left to bottom right. Limestone. The hammer is ~30 cm long. B: *Hummocky forms with steep lee faces* developing from coalescing *potholes*. In the left of the picture, a complete *steep lee face* can be seen, while on the right is a line of partially coalesced *potholes*, which will ultimately develop into a *steep lee face*. Gayle Beck, UK. Flow from right to left. Limestone. The scale is 60 cm long.

reduces the amount of flow separation in the lee of an obstacle to the flow, reducing its drag (Richardson, 1968), and represents streamlining of the rear of the *upstream-facing convex surface*.

4. Solutional Forms

This is the fourth major class of forms and is a genetic classification. It consists of bedforms that develop almost exclusively in limestone and, by inference, principally by the mechanism of dissolution. Some, especially *scallops*, are poorly developed in other, less soluble rocks, such as quartzites (Ford and Williams, 1991: p. 27–29), granite, and volcanic rocks rich in glass. Included within the group are *solution pits* and *pans*, *scallops*, and a range of other *solutional forms*.

4.1 Solution Pits and Pans

An apparently simple group of forms, *solution pits* are in fact diverse morphologically and occur in diverse environments. They represent sites of locally enhanced dissolution. *Solution pits* are small depressions, having dimensions of millimeters to a few centimeters. This group contains equidimensional, elongated, and curvilinear forms in arrangements ranging from fully isolate through space-filling to coalesced arrangements (Fig. 113). The axes of symmetry or of elongation of these forms are parallel to flow. *Solution pits* are also instrumental in the formation of other, larger bedforms such as *potholes*, *flutes*, *convex surfaces with steep lee faces*, and *scallops*. In calcareous rocks, they sometimes act as an initial locus of attack upon an otherwise smooth bed, generating defects and flow separation, and thus opening the bed up to more rapid erosion, as the example of *steep lee faces* (section 3.3a) and examples below illustrate. The fundamental division of *solution pits* is into *fluvial* and *nonfluvial* types. It is common for both types to be present at a site, in different parts of the channel.

(a) *Fluvial solution pits.* *Solution pits* are common in areas of limestone channels that, although not permanently wetted, are nonetheless close to the thalweg, and these areas are heavily sculpted by a variety of fluvial bedforms. In view of their environment and morphology, these *solution pits* appear to be true fluvial *solutional forms*, although it is possible that they are karst forms that survive substantial periods of fluvial erosion and that are strongly modified by the flow. In either case, these *solution pits* experience significant streamflow, which acts to produce smooth internal walls and surrounding external rock surface, and rounded rims to the *pit*, at least in part. The other important effect of a strong fluvial influence is to produce *solution pits* elongated in the direction of flow. Therefore the circular types, which are at one end of the spectrum of elongation, occur on surfaces not experiencing a strong unidirectional flow,

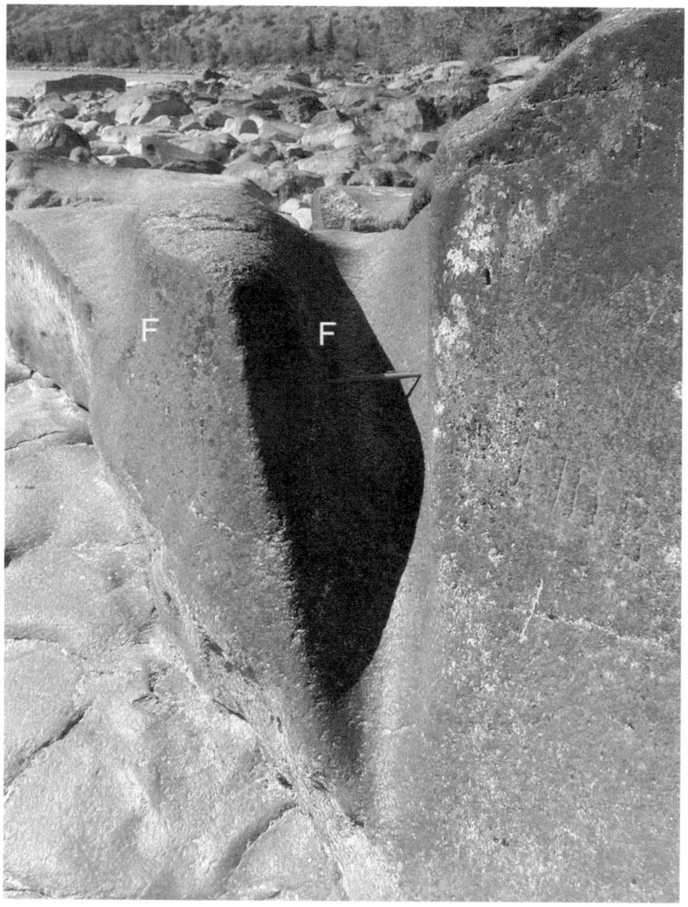

Figure 112. *Streamlined steep lee face*. Note the *furrows* (F) giving the rock projection an angular, downstream-pointing lee side where they terminate, and producing a sharp vertical crest, which is analogous to a tail fin, and which represents the streamlining of the feature. Katun River, Siberia, looking upstream. Pen for scale.

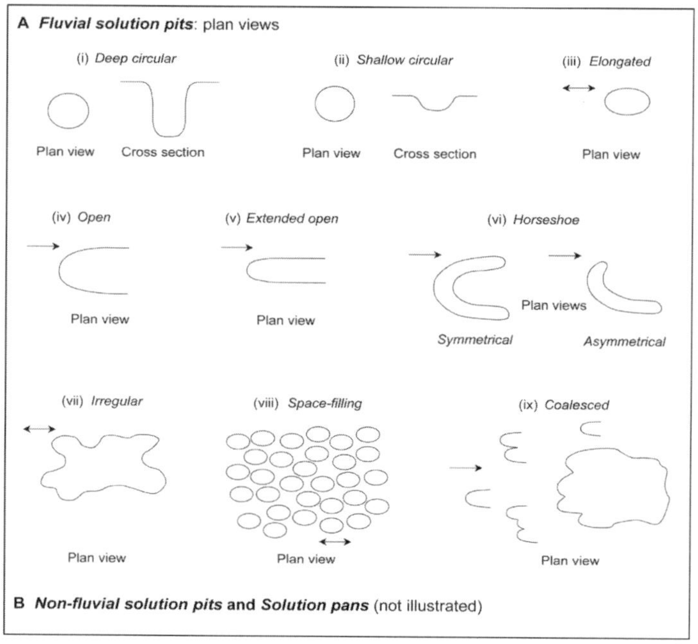

Figure 113. *Solution pits*. Arrows indicate flow directions.

such as areas of relatively slack water and areas within the flow separation cells generated by larger sculpted forms or bedrock projections. In other areas, elongated forms prevail. Types (ii) to (v) below form a sequence of increasing length:width and length: depth ratios. Types (ii) to (iv) and (vi) can form on any surface but occur preferentially in areas of flow expansion and flow separation, such as the lee faces of *potholes*, *hummocky forms* and other irregularities. It seems likely that the adverse pressure gradients and/or flow separation developed in these environments are conducive to their formation. However, if these *solution pits* are not true fluvial sculpted forms, it is also possible that where they are found only on the lee slopes of other bedforms, it is because they are removed from the stoss sides as a result of the higher rate of fluvial erosion there. There is a continuum of form between *solution pits* of types (ii) to (vii) listed below, and a continuum in size between *solution pits* and *potholes*, which occur on a decimeter to meter scale. Also, with decreasing fluvial influence, either due to the stagnation of the overlying water or due to infrequent inundation, the smooth *fluvial solution pits* grade into the rough *nonfluvial solution pits*.

(i) *Deep circular solution pit*: These are cylindrical shafts that are deeper than they are wide, several centimeters across with vertical or near-vertical walls, yet too small to be called *potholes* (Fig. 114). They are curious features that were not observed to form a continuum with the other types of *solution pits*, although this may be because of the small number of examples discovered. They are developed in isolation from other bedforms on *concordant plane beds*.

(ii) *Shallow circular solution pit*: These are *circular pits* with a rounded floor that are not as deep as they are wide (Fig. 115). The slope of the walls is variable but typically has a maximum value of 20°–40° in *solution pits* measured in the River Dee. The transition to *pothole* status is accomplished not only by an increase in size, but also by an increase in relative depth and the slope of the walls.

(iii) *Elongated solution pit*: The usual form of *fluvial solution pits* is to be slightly elongated (Fig. 116). This occurs through preferential erosion of the downstream wall of the *pit*, resulting in an asymmetrical *solution pit* that is steeper on the upstream wall than on the downstream wall, the deepest point occurring nearer to the upstream than the downstream rim. The downstream rim experiences wear and is highly rounded in cross section. In plan view, the rim on the upstream side remains semicircular.

(iv) *Open solution pit*: With increasing amounts of erosion focused on the downstream wall, the downstream part of the rim becomes diffuse. The slope of the downstream wall is now very low, and the floor of the *solution pit* merges into the surrounding rock surface on that side, so that the rim is no longer continuous around the *solution pit*. The *pit* is therefore open on the downstream side (Fig. 117). The upstream part of the rim may, however, be cuspate. This type of *solution pit* is sometimes similar in appearance to a small *flute*. It is classed as a type of *solution pit* rather than a *flute*, however, because there is no obvious defect at which it originated, and because it forms a continuum with the other types of *solution pits* described. Furthermore, the rim on the upstream side of the *pit* remains approximately semicircular (rather than parabolic) in plan view, in common with the other types of *solution pits* (except the *irregular* variety; see below). Both *elongated* and *open solution pits* sometimes form in rows orientated diagonally with respect to the flow direction.

(v) *Extended open solution pit*: In high-gradient or constricted reaches where flow velocity is high, *open solution pits* may be found that are several times longer than wide (Fig.

Figure 114. *Deep circular solution pit*. This feature is not considered a *pothole*, owing to its small size. Nam Mae Chaem (Ob Luang), Thailand. Granitic gneiss.

Figure 115. *Shallow circular solution pits*. River Dee, UK. Flow from top to bottom. Limestone. The hammer is 30 cm long.

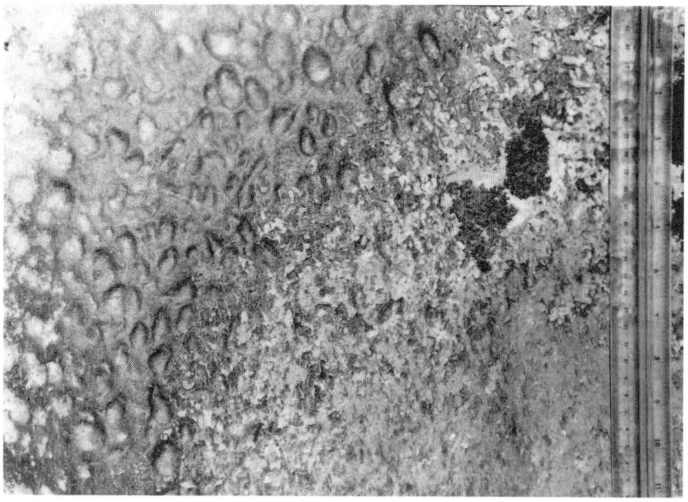

Figure 116. *Elongated solution pits*. The *pits* are elongated in the direction of flow. River Dee, UK. Flow from bottom to top. Limestone. The left-hand scale on the ruler is in centimeters.

Figure 117. *Open solution pits* (in places coalescing). River Dee, UK. Flow from top to bottom. Limestone. Keys for scale.

118). The upstream part of the *solution pit* is similar to the short open type, but in the downstream section, the two limbs of the rim run parallel for some distance, before merging with the surrounding rock surface. These *solution pits* do not have the pointed shape (in plan view), or length:width ratio, of a *spindle-shaped flute*, however.

(vi) *Horseshoe solution pit*: This is a confusing bedform when first encountered because it has two varieties: symmetrical and asymmetrical. These features were found only in the River Dee, but at several sites hundreds of meters apart. They are small *solution pits*, being generally ~10 mm wide, with a length only slightly in excess of their width. Thirty examples of the symmetrical variety were measured at one site, and these showed a remarkable uniformity in width, with a mean value of 9.5 mm and a 95% confidence interval of ±1.6 mm. The length of these features was more variable, however. They occur in turbulent environments, immediately downstream of both *forward-* and *backward-facing steps*. They occur not in isolation, but in closely packed groups (see below). They also occur on the lee faces of *hummocky forms* associated with *elongated* and *open solution pits*. The symmetrical variety is a small horseshoe-shaped *furrow*, the open part of the horseshoe pointing downstream (Fig. 119A). It is analogous to both a *flute with median ridge* and the s-forms sichelwannen (Kor et al., 1991), but on a rather different scale. In the asymmetrical variety (Fig. 119B), one limb of the horseshoe is absent or poorly developed, which is analogous to an *asymmetrical obstacle mark* and the s-forms known as comma forms. The result is a small hook-shaped or crescentic *furrow* subparallel to flow, whereas the symmetrical forms are transverse crescents. When both varieties are present together, the appearance may at first therefore be of randomly orientated curvilinear *solution pits*.

(vii) *Irregular solution pit*: Some *solution pits* have an irregular plan outline and internal morphology (Fig. 120). They vary from equidimensional pits to elongate hollows parallel to flow. Those seen by the authors vary in size from 1 to 40 cm in diameter and up to 15 cm in depth. In the larger *pits*, a hierarchical (i.e., compound) structure may exist, with smaller *pits* developing within them. Internal rock structure is important in the production of these bedforms. Rocks with a heterogeneous structure at the centimeter to decimeter scale will produce mainly *irregular solution pits* rather than types (i) to (vi). They may also be favored by lamination, thin bedding and other thinly

Figure 118. *Extended open solution pits*. River Dee, UK. Flow from left to right. Limestone. Penknife for scale.

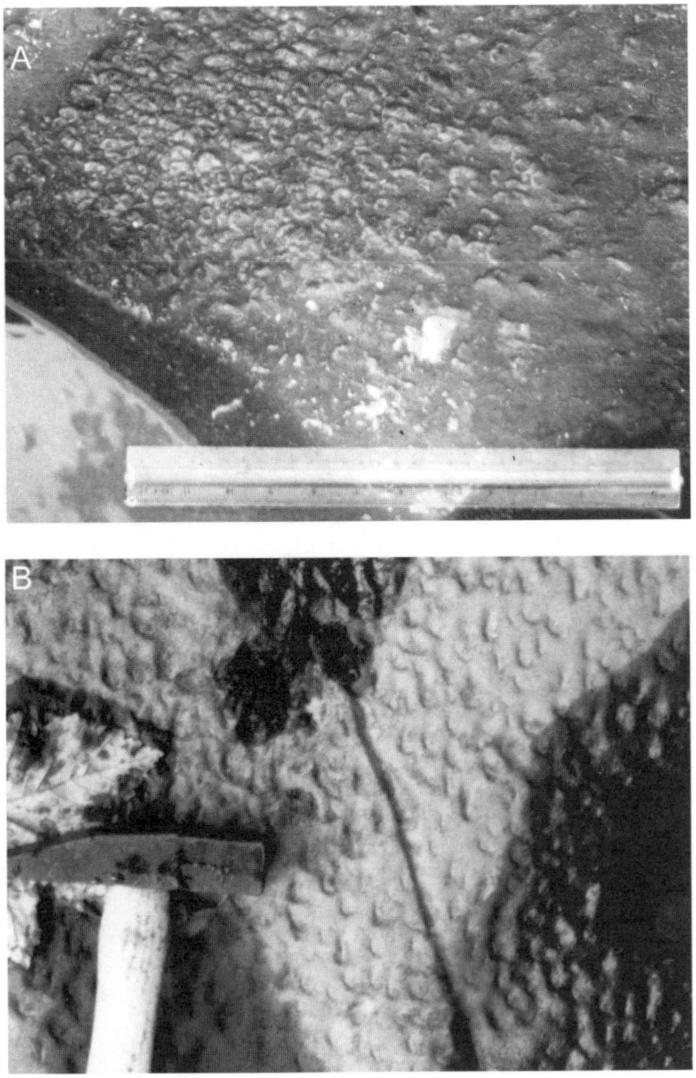

Figure 119. A: *Horseshoe solution pits*, mostly of the *symmetrical* variety, occurring as a space-filling assemblage. River Dee, UK. Flow from top left to bottom right. Limestone. The ruler is 30 cm long. B: *Horseshoe solution pits* of the *asymmetrical* variety. River Dee, UK. Flow from bottom to top. Limestone. Hammer for scale.

layered structures, because at the Sleightholme Beck and River Lune sites, the internal morphology of *irregular solution pits* reflects the layered rock structure present. Layered structures promote undercutting of the rims, where a less resistant layer may be exploited below the surface layer. Very high flow speeds may also be important in the development of *irregular solution pits*. At Sleightholme Beck, for instance, *irregular solution pits* occur only in the area of accelerating flow immediately above a waterfall, despite the widespread occurrence of heterogeneous and layered rock in the streambed. In these examples, the downstream ends of the *pits* are the loci of the most intense erosion. The downstream ends are the deepest and are often undercut.

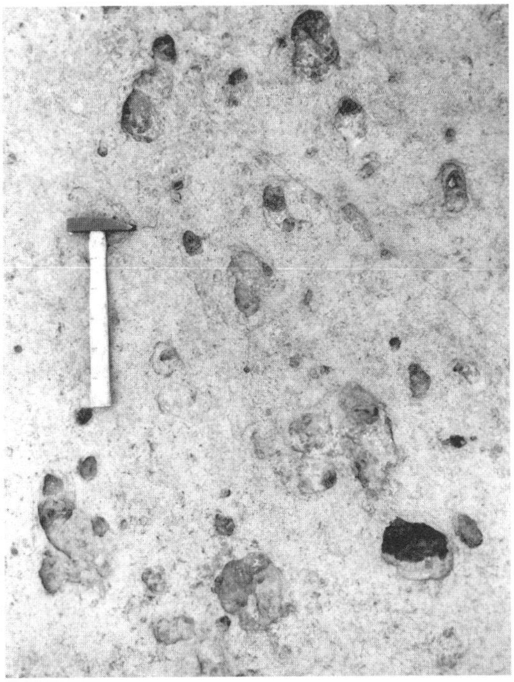

Figure 120. *Irregular solution pits*. Flow from bottom right to top left. Sleightholme Beck, UK. Limestone. The hammer is 30 cm long.

(viii) *Space-filling solution pits*: This arrangement is uncommon and was observed only in restricted environments in the River Dee. Its essential characteristic is a densely packed, space-filling arrangement of the *solution pits*, despite which individuals are isolate and are not interfering (Fig. 121). *Elongated*, *open*, and *horseshoe solution pits* may be found in this type of assemblage. The restricted environments in which it may be found are highly turbulent ones, for example, immediately downstream of *steps*, both *forward-* and *backward-facing*. Within these areas, *space-filling solution pits* are found to occur preferentially on the lee faces of *hummocky forms* (Fig. 121). More or less regular patterns of *space-filling solution pits* may be found. For instance at one site that has a *concordant plane bed*, *space-filling horseshoe solution pits* occur in wide bands orientated diagonally to the flow. The bands are separated by unpitted rock, and the spacing of the bands is much greater than the dimensions of individual *solution pits*. At this site it was observed that in moderate flows, the area of the diagonal bands was occupied by oblique shock waves, which may therefore be related. Another pattern that is often found in space-filling assemblages of any type of *solution pit* is the tendency for *pits* to form in diagonally orientated rows, although this type of order does not extend farther than a handful of individuals in any one row. Such arrangements may be called en echelon and are also present in some assemblages of the s-forms sichelwannen (Kor et al., 1991; Shaw, 1996) and in rhomboid ripples (Hoyt and Henry, 1963; Otvos, 1965). Longitudinal rows also occur but are less common. In places, *solution pits* in space-filling assemblages do begin to interfere and become partially conjugate. In such cases, they begin to resemble incipient *three-dimensional scallops* (see below), and may be transitional with them. This will be discussed further in a later section.

(ix) *Coalesced solution pits*: Just as *solution pits* may coalesce to form *steep lee faces* on *hummocky forms*, in relatively flat areas of intense pitting they may coalesce to form wide, shallow depressions of irregular outline and internal morphology. This is especially true of *open solution pits* (Figs. 117 and 122). In *coalesced solution pits*, individuals eventually lose their identity and are no longer discernible.

(b) *Nonfluvial solution pits* and *pans*. In areas of the channel lacking a strong or frequent fluvial influence, such as the higher parts of the channel margins and areas that often contain very slow moving or stagnant water, *solution pits* and *pans* form through the action of standing rather than flowing water (e.g., King, 1927). Such *nonfluvial solution pits* in limestone are strictly karren features but grade into their fluvial counterparts with an increasing fluvial influence.

(i) *Solution pits*: These are circular or irregular hollows with diameters of centimeters to decimeters, with rounded floors (Ford and Williams, 1991: p. 379) and no tendency for elongation parallel to flow in the channel (Fig. 123A). The rims are cuspate, and the rock surface inside and outside the *pit* is rough and/or irregular, with no sign of polishing by the flow of water. *Nonfluvial solution pits* often appear in conjugate assemblages (Fig. 123B). Such surfaces are composed of steep-sided, cup-like depressions of various sizes separated by sharp, relatively delicate and serrated

Figure 121. *Space-filling open solution pits*. River Dee, UK. Flow from top to bottom. Limestone. The hammer is 30 cm long.

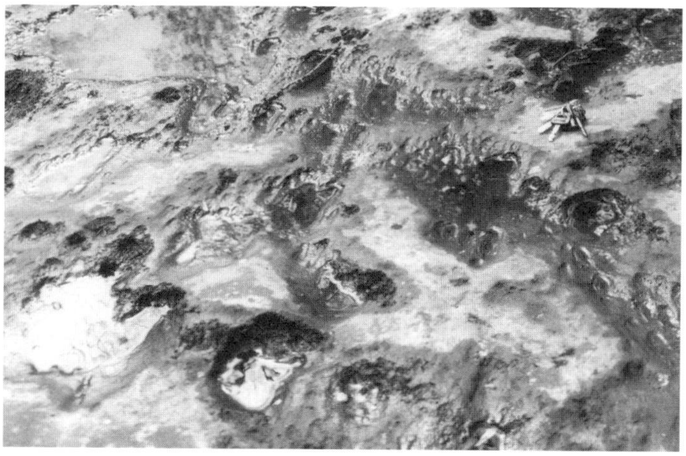

Figure 122. *Coalesced fluvial solution pits* (the wider, irregular depressions) and some remaining isolate *solution pits*. River Dee, UK. Flow from top right to lower left. Limestone. Keys for scale.

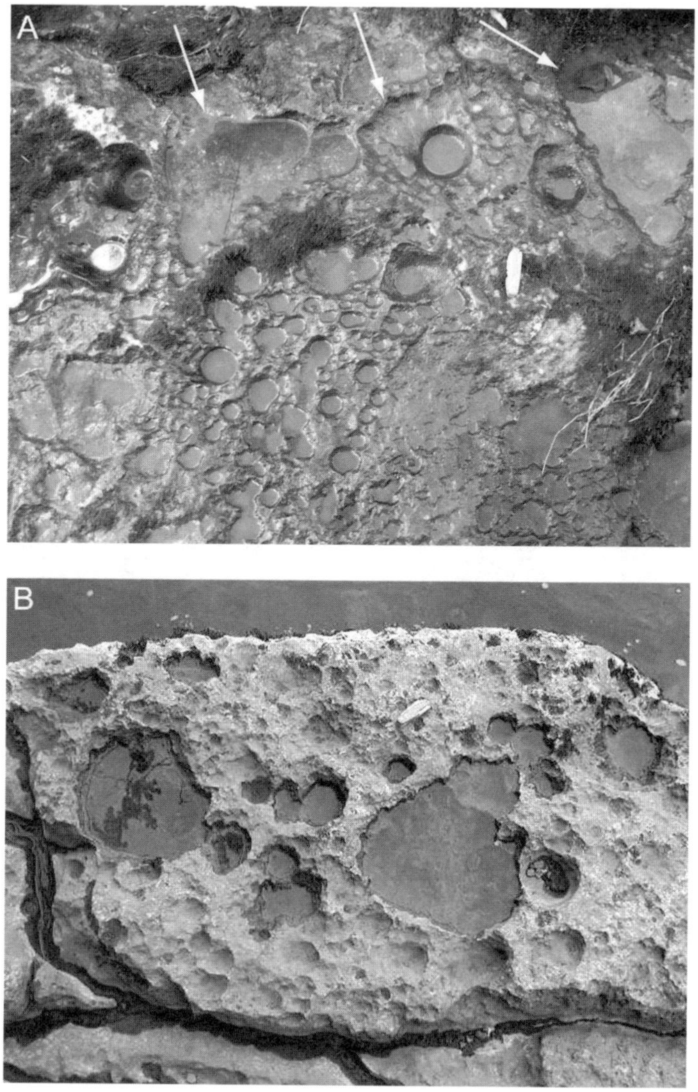

Figure 123. A: *Nonfluvial solution pits*. The *pits* have sharp rims and are radially symmetrical. In places, the *solution pits* have coalesced and are forming *solution pans* (arrowed). River Kent, UK. Limestone. Penknife for scale. B: Conjugate *nonfluvial solution pits* are creating a surface with a cindery appearance, often referred to as "cockling." River Wharfe, UK. Limestone. Penknife for scale.

crests, or occasionally steep but flat-topped crests. The rock has a crinkled or cindery appearance, which is sometimes referred to as "cockling" (Allen, 1982: p. 242).

(ii) *Solution pans*: These are larger features than *solution pits* (decimeter to meter scale) but are not necessarily any deeper and have near-vertical walls and flat floors (Ford and Williams, 1991: p. 379). They form via the enlargement of individual *(nonfluvial) pits* or by the coalescence of several of them (Fig. 123A), which produces an irregular and cuspate rim. Once established, they may contain standing water for long periods of time and become colonized by algae, which could potentially have a positive feedback effect by firstly inhibiting the drying out of the *solution pan* and secondly releasing carbon dioxide from decaying organic matter and reducing the pH of the water.

4.2 Scallops

Scallops are trains of contiguous depressions delimited by sharp crests that in general trend in the transverse direction (Fig. 124). Although they are well known from caves, where they occur as phreatic features (i.e., developed below the water table; Gale, 1984), well-developed examples are not common in surface channels, from which they have not previously been reported. Following the work of Allen (1971a, 1971b) and Goodchild and Ford (1971) and observations made for this report, it is hypothesized that all *scallops* are generated from *flutes* and/or *solution pits* of various kinds by a two-step process involving (1) the complete conjugation of initially isolate forms and (2) their subsequent modification by mutual interaction under the prevailing conditions, resulting in an assemblage that is adjusted to the hydraulic regime (see Discussion). The result of this process is that the character of the original isolate *flutes* or *solution pits* is considerably changed, as discussed in the terminology section and in section 1.2a, and in becoming *scallops* the pattern takes on the appearance either of ripples or of equidimensional, polygonal depressions.

Previous descriptions of *scallops* are insufficiently descriptive, or even cryptic. For example, *scallops* are described as "interrupted concavities" (Curl, 1966), "ripple-like features ... [resembling] a mosaic of inlaid scallop shells" (Goodchild and Ford, 1971), "polygonal, intersecting depressions" (Gale, 1984), and "spoon-shaped scoops" (Ford and Williams, 1991: p. 303). Such descriptions are also too narrow and fail accurately to represent the wide spectrum of forms that exists by describing only its end members. Most forms observed in this study are in the midrange between these end members and bear little resemblance to the majority of published descriptions. Allen (1971a), in his seminal work on *flutes* for instance, considers *scallops* as being simply conjugate *flutes* and describes them according to his scheme for *flutes*. In the view of the authors, this is an oversimplification in the light of both the two-step process proposed above and the observations of morphology made for this report. Consequently, Allen (1971a) fails to do justice to the description of *scallops*, particularly with regard to crest planform, which he describes as being mainly "polygonal." The classifications below are based on crest symmetry and crest planforms. The fundamental division of *scallops* is into *nondirectional* and *directional* forms.

(a) *Nondirectional scallops*. These are closed depressions with no clear orientation. In planform, they are equidimensional polygons defined by gently curved, sharp-crested ridges that are symmetrical or at least show no systematic asymmetry (Fig. 125).

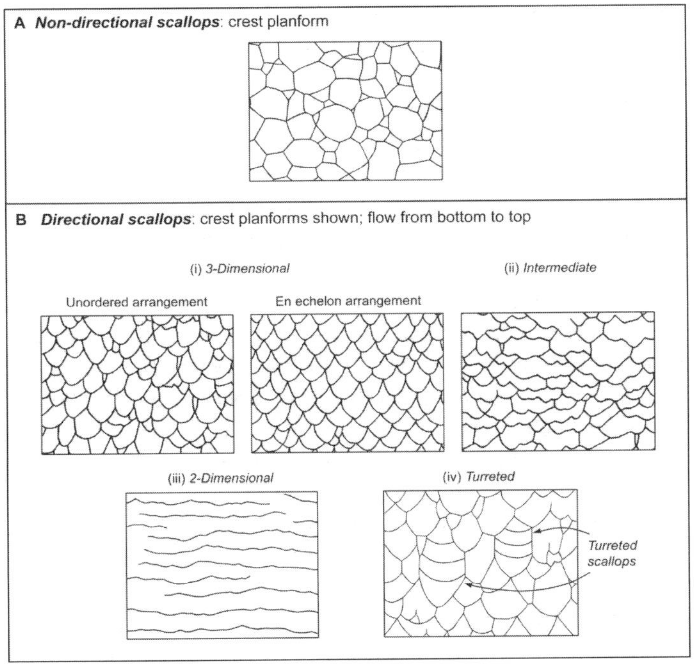

Figure 124. *Scallops*.

Figure 125. *Nondirectional scallops*. These *scallops* show no clear orientation. The surface on which they occur faces upstream. Sleightholme Beck, UK. Limestone. Penknife for scale.

There is a wide range of *scallop* sizes within a pattern because of space-filling requirements. These bedforms are analogous to polygonal karst. As with *circular solution pits*, which also lack a clear orientation, they form on surfaces on which a strong unidirectional flow is absent. There are many types of such surfaces, leading to a spectrum of forms being developed, depending on the hydraulic environment. A strong unidirectional flow may be absent because the bedrock surface is in relatively slack water, in which case the *scallops* are relatively deep and steep-sided depressions that grade into the cindery appearance of conjugate *nonfluvial solution pits*. Other surfaces are not in slack water but still do not experience a strong unidirectional flow; these include the vertical sides of boulders and bed irregularities that face directly upstream or downstream, surfaces experiencing the normal impact of a jet of water flowing over a *step*, surfaces within flow separation cells, and surfaces experiencing the reattachment of a separated boundary layer. In these cases, the *scallops* occur as shallower depressions with relatively low-angle (normally 15°–30°) crests and a greater degree of smoothing of the surface compared with *scallops* forming in slack water. These different types grade into each other, sometimes at a single site as one moves between different environments. *Nondirectional fluvial scallops* appear to arise in general through the development of densely packed assemblages of *circular fluvial solution pits* that become conjugate. In some cases, this transition can be observed in space as *circular solution pits* pass into *nondirectional scallops*. This category of *scallops* includes the "undulation flutes" of Maxson and Campbell (1935).

(b) *Directional scallops*. These *scallops* have a ripple-like appearance, with asymmetrical crests that are continuous in the transverse direction to a greater or lesser extent. The stoss slopes are slightly convex, while the steeper lee slopes are slightly concave. This asymmetry gives the bedforms a clear orientation that can be used as an indicator of paleoflow direction (Coleman, 1949) and is a result of their formation under strong unidirectional flow. Furthermore, the wavelength of *directional scallops* has been found to be inversely proportional to current speed (Curl, 1966; Allen, 1971a; Goodchild and Ford, 1971; Blumberg and Curl, 1974) and can be used to infer paleoflow velocity (Curl, 1974; Lauritzen et al., 1983; Gale, 1984; Murphy et al., 2000; Springer and Wohl, 2002). In the fieldwork undertaken for this study, extensive examples of *directional scallops* were found only at four sites, although at one site (Sleightholme Beck) they are all but ubiquitous. In all four sites, the channels are steep and incised (i.e., they are small gorges) or are in a steepened, rapidly incising region immediately above a *knickpoint*, and it seems that, in contrast to cave passages, the development of extensively scalloped surfaces in open channels requires high-velocity and/or highly turbulent flow. Elsewhere, *scallops* are only poorly developed. These conditions are reflected in the small wavelengths of *directional scallops* in surface channels. Wavelengths observed during this study varied from 1 to 10 cm, compared with the range of 5–20 cm given by Allen (1971a) for *scallops* in caves. *Scallop* wavelengths were found to be short (typically 2–4 cm) in the high-gradient reaches where they are best developed, and longer (5–10 cm) in lower gradient reaches, where the *scallops* were only poorly developed or incipient.

Certain types of concave s-forms with diameters up to 2 m are sometimes referred to as "scallops" (e.g., Glasser and Nicholson, 1998; Bradwell, 2005). On the basis of their supposed morphological similarity to *scallops* in caves, such s-forms are interpreted as phreatic features (Glasser and Nicholson, 1998). However, in the view of the authors, this is a misidentification. Firstly, their large wavelength implies very slow mean flow speeds on the order of a few millimeters per second. It does not seem possible for such slow currents to entrain significant amounts of suspended load and to erode large bedforms. Secondly, on morphological grounds, these s-forms cannot realistically be compared to *scallops* because they are not fully conjugate and do not form extensive trains of bedforms. It appears that these s-forms would be better described as *flutes*. This example emphasizes the need, as discussed in the Terminology section in the context of *potholes*, to adhere to strict definitions of bedforms and not to widen the application of bedform names.

As might be expected, there is a spectrum of forms between *nondirectional* and *directional scallops* that matches the range of flow conditions experienced by rock surfaces. This indicates that all types of *scallops* are closely related, with similar origins; hence the hypothesis outlined at the start of section 4.2. As one moves from an environment without strong unidirectional flow to one with such a flow, a corresponding change in *scallop* morphology is achieved by an increasing asymmetry of the crest slopes, by an increasing number of crest sections being orientated transverse to the flow, and by the *scallops* developing a characteristic wavelength. In addition, in plan view, the crests acquire parabolic sections that are convex-upstream, and angular sections that point downstream and end in a peak at the apex.

There is also a gradation within *directional scallops* between *three-* and *two-dimensional* forms. The classification below is based on this aspect of their morphology. *Scallop* assemblages consisting purely of one of the types listed below are rare or nonexistent, at least over large areas. In nature, assemblages generally consist of mixtures of them, especially of types (i) and (ii), and may be classified according to which type is dominant. Transitions between types of assemblages occur in a subtle manner either by collective changes in the character of the bedforms or by changes in the proportions of the different types present in the assemblage, or both. Detectable changes in the character of *scallop* assemblages can occur over distances of just a few decimeters.

Planform patterns of *scallop* assemblages are shown in Figure 126, and longitudinal profiles in Figure 127. The line drawings of crest planform patterns are used merely to illustrate examples of *three-dimensional* and less *three-dimensional* (i.e., *intermediate*) *scallops*. It should be borne in mind that line drawings of planform patterns are a crude tool and give a poor impression of the true appearance of *scallops*. This is because they do not show the high and low points, or which crests are the steepest or most prominent, and because they do not show the three-dimensional shapes (i.e., contours) of the depressions and ridges. Further, the reduction of a *scallop* assemblage, which is a highly three-dimensional surface, to a two-dimensional network of crest sections, and its analysis as a network of nodes and lines (e.g., Allen, 1971a: p. 177–178), yield little insight. This is because so much information is contained within the shape of the surface and, in particular, in the spatial relationships between peaks and *troughs*. It is more useful to talk of different types of peaks than it is to talk of types of nodes, as will be discussed below. The "broad pocket flutes" of Maxson and Campbell (1935) are in fact a variety of *directional scallops*. Dzulynski and Sanders (1962: Plates IVB and VA therein) describe transverse ripple-like scours ranging from two- to three-dimensional in character and similar to *directional scallops*, formed in mud and preserved in the geological record.

(i) *Three-dimensional*: These *scallops* are the spoon- or cup-shaped hollows, interrupted concavities, and "inlaid scallop shells" referred to by many authors (e.g., Curl, 1966; Goodchild and Ford, 1971; Ford and Williams, 1991: p. 303; Slabe, 1996: p. 15). The *three-dimensional scallop* planform pattern consists of roughly equidimensional hollows defined by a network of short crest sections. The two important features of the crests in plan view are the dominance of convex-

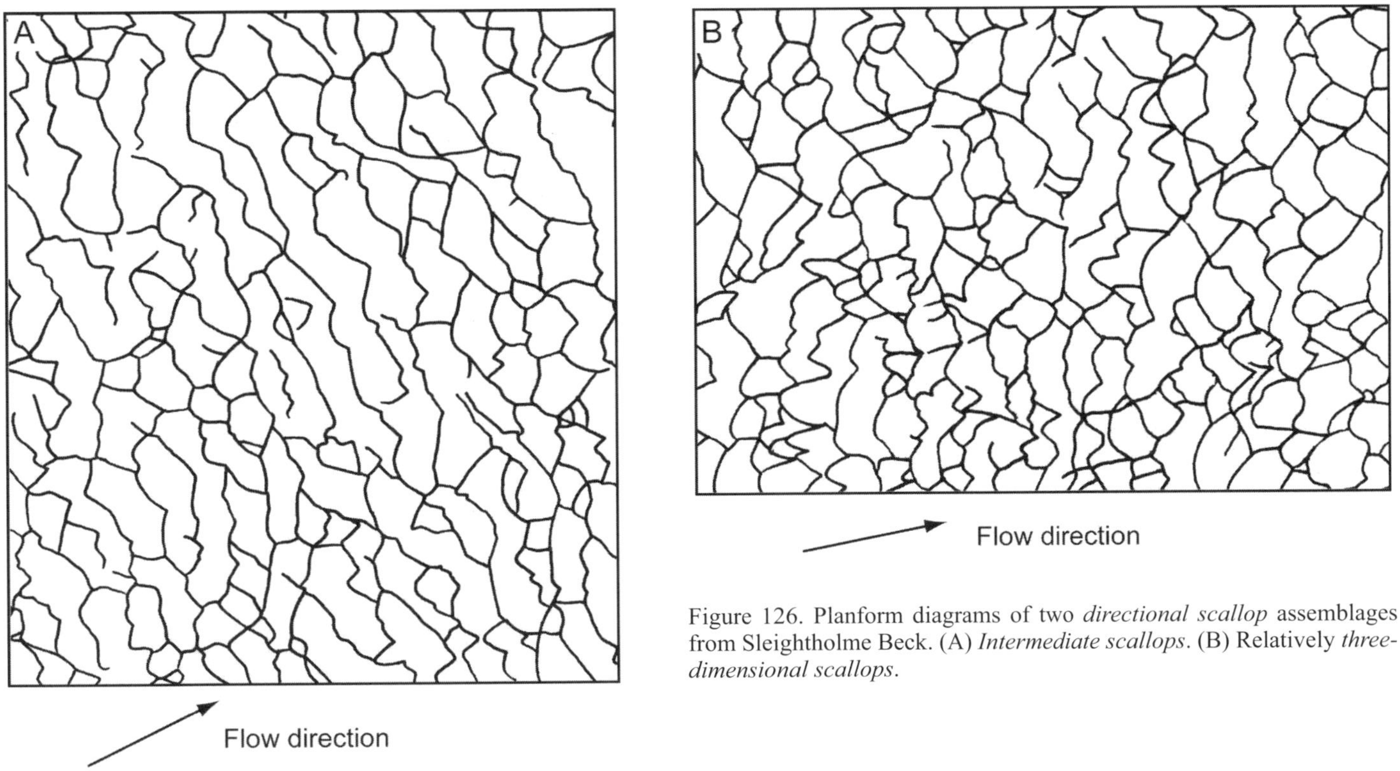

Figure 126. Planform diagrams of two *directional scallop* assemblages from Sleightholme Beck. (A) *Intermediate scallops*. (B) Relatively *three-dimensional scallops*.

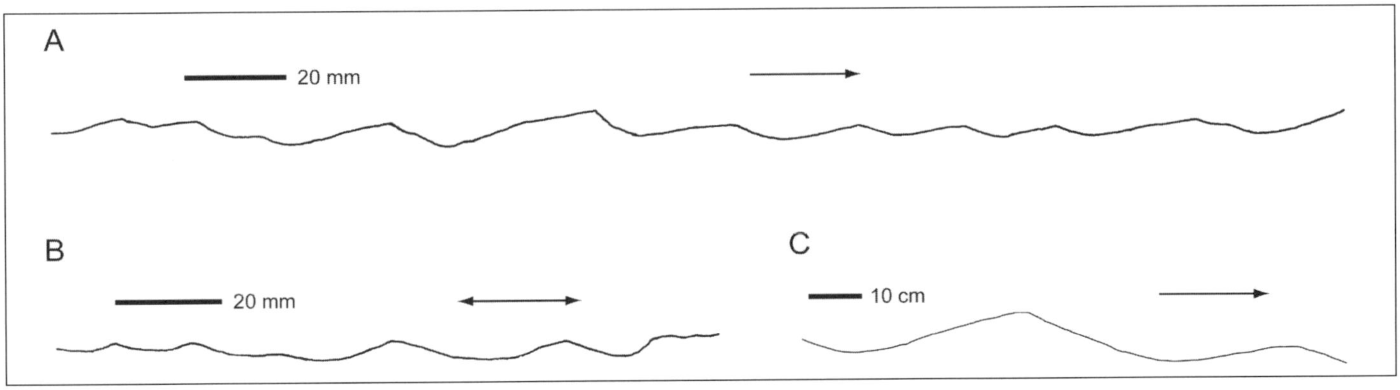

Figure 127. Longitudinal profiles of ripple-like features. A: *Directional scallops*. B: *Nondirectional scallops* (note the crest symmetry). C: *Sharp-crested hummocky forms*. Arrows indicate direction of flow.

upstream, parabolic crest sections and downstream-pointing, angular sections (Fig. 128). The parabolic crest sections enclose hollows that are close in shape to the proximal portions of the parabolic scour hollows *flutes*. The downstream-pointing angular sections slope upwards, ending in a peak at the apex, and are generally formed by the presence of two parabolic crest sections lying side by side. The peak typically has a longitudinal crest section on its downstream side. These two elements (the parabolic sections and the angular sections) are combined in the "ideal" *three-dimensional scallop*, which has a parabolic upstream crest enclosing a hollow, and a triangular downstream region ending in a peak (Slabe, 1996: p. 31). Such *scallops* are indeed common, but in any real assemblage, shapes are variable and other types will also exist. Coleman (1949) described how *three-dimensional scallops* have a significantly different appearance depending upon the direction from which they are lit. When lit from the upstream direction, the appearance is that of an array of triangular peaks, while lit from the downstream direction, the curved upstream crests are picked out and the appearance is that of ripples. It should be noted that, as a consequence of the three-dimensionality of the pattern, crest sections are generally short and crests show a high degree of connectivity in both transverse and longitudinal directions. In the experience of the authors, *scallops* always lack a *median ridge.* The mature, fully conjugate experimental *flute* assemblages of Allen (1971a: Figs. 70C, 70D, and 71 therein) are pure assemblages of *three-dimensional scallops* according to the present classification.

A common, but by no means dominant, arrangement of *three-dimensional scallops* is an ordered one involving the ideal form and is here termed the en echelon arrangement. In *en echelon scallops*, successive individuals are in a more or less regular antiphase arrangement, in which the apex of a downstream-pointing angular crest section approaches the upstream end of the parabolic crest of the next *scallop* downstream (Fig. 129). Diagonal alignments of both depressions and crest sections are thereby created. This arrangement is similar to the en echelon assemblages of *flutes* and *space-filling solution pits* noted in sections 1.2a and 4.1a. Such patterns exist only over short distances, say not more than ten individuals in any direction. An antiphase arrangement with diagonal alignments of *three-dimensional scallops* was reproduced in some of the experiments of Allen (1971a: e.g., Fig. 70C) but was not remarked upon. En echelon arrangements of s-forms (sichelwannen) also occur (Kor et al., 1991; Shaw, 1996), as do antiphase arrangements of the sedimentary structures known as lunate or barchanoid ripples (Allen, 1968) and the rhomboid ripples of Hoyt and Henry (1963) and Otvos (1965). In-phase (i.e., longitudinally aligned) arrangements of *three-dimensional scallops* also occur but are less common.

The peaks within a *three-dimensional scallop* assemblage, which are formed at the apices of the downstream-pointing angular crest sections, are of six types, as illustrated in Figure 130. The most common are types A and B, the least common is type F, and the remainder are approximately equally common. Of the six, only types B–D can lead, if repeated, to en echelon assemblages. Types A–D form a sequence of decreasing distance between the peak and the parabolic crest section of the next depression downstream. In type A they are separated by a long ridge, gently sloping or horizontal over most of its length and not necessarily

Figure 128. An assemblage dominated by *three-dimensional scallops*. Note the tendency for crest lines to form downstream-pointing triangular regions and convex-upstream parabolic regions. Sleightholme Beck, UK. Flow from bottom to top. Limestone.

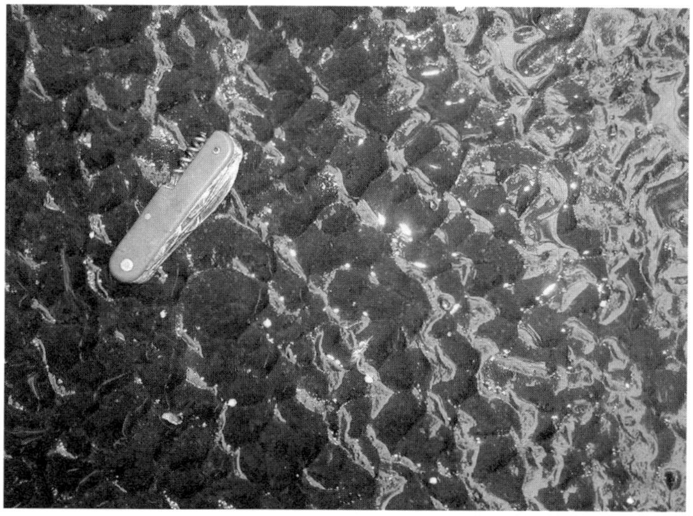

Figure 129. An assemblage dominated by *three-dimensional scallops*. This assemblage has a strong en echelon tendency. Sleightholme Beck, UK. Flow from bottom to top. Limestone.

straight, and in type B by a short, straight ridge sloping steeply down to the depression. In type C they are in contact, while in type D the depression cuts into and truncates the peak. Type E shows a hierarchical arrangement of crest sections, while type F is an in-phase arrangement of peaks. Peaks of type A are apparently subject to an instability in which the peak acts as a defect and eventually nucleates a new *scallop* immediately downstream of itself.

(ii) *Intermediate*: *Intermediate scallops* differ from *three-dimensional scallops* by the development of straighter and longer transversely trending crest sections that are more dominant over the longitudinally trending crest sections (Fig. 131). This change is due to the fact that, although still common, there are fewer angular downstream-pointing crest sections (and therefore fewer peaks), and many of those that remain are less angular. There are also fewer curved convex-upstream portions within the crests, and correspondingly fewer parabolic scour hollows. The decrease in the number of angular crest sections means that there are fewer longitudinal crest sections, since these are mainly developed as ridges on the downstream sides of peaks. Furthermore, of the remaining longitudinal crest sections, many are more diffuse and simply fade out without joining another crest. Collectively, these changes result in more dominant, longer, and straighter transverse crests and a lower connectivity between crests in the longitudinal direction. Thus, the crests of *intermediate scallops* are now more two-dimensional and are similar in appearance to sinuous ripples. However, an interesting feature of *intermediate scallop* assemblages is the presence of a subordinate, but often strong trend for long, diagonal crest lines crossing several individual *scallops*. This suggests an evolution of *intermediate scallops* from *three-dimensional* ones, whereby the en echelon tendency is preserved in diagonal crest lines.

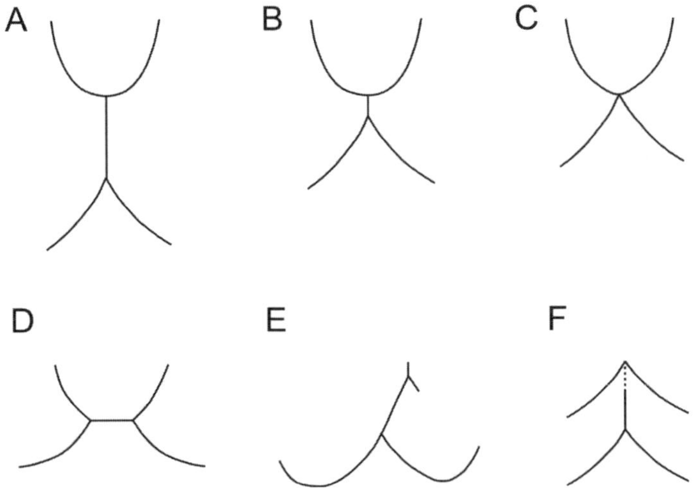

Figure 130. The six different types of peaks identified in *three-dimensional scallop* assemblages. The dotted line represents a weak or sometimes absent ridge. Flow direction is from bottom to top. See text for explanation.

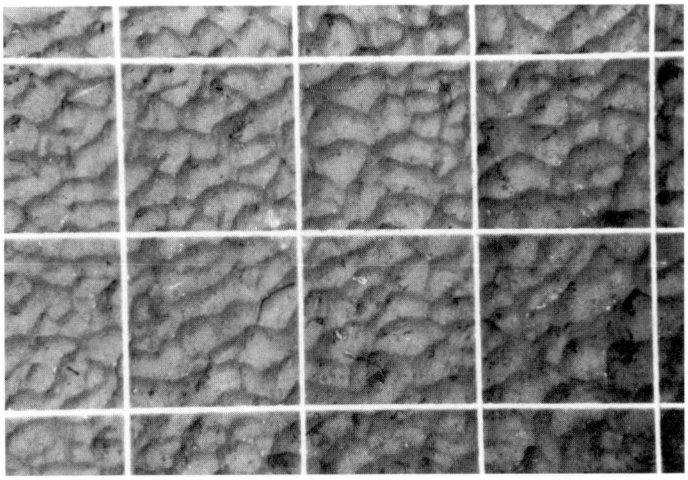
Figure 131. An assemblage dominated by *intermediate scallops*. Sleightholme Beck, UK. Flow from top left to bottom center. Limestone. The grid squares are 10 cm apart.

Individual *scallops* are generally elongated in a transverse direction as a result of the loss of longitudinal crest sections that divided laterally adjacent individuals. As one might expect from this increase in width, the *scallops* now often contain two or more low points, which may or may not be *flute*-like parabolic scour hollows, and which are separated by gently rounded and subdued ridges. They are effectively *three-dimensional scallops* that have in places coalesced in the transverse direction. At Sleightholme Beck, most *scallop* assemblages are dominated by *intermediate scallops*. The *scallops* illustrated by Springer and Wohl (2002: Fig. 1A therein) would also probably fall into this category.

(iii) *Two-dimensional*: These are the two-dimensional end member of the series. They are the features formerly referred to as cave "flutes" by many authors. All longitudinal contact between crests has been lost, and crest sinuosity is low. There are few, if any, downstream-pointing angular crest sections and parabolic crest sections. Crests are long and orientated almost universally in the transverse direction (Fig. 132). Individual *scallops* (i.e., the depressions) are therefore of great lateral extent and are poorly defined (being no longer closed depressions). Such *scallops* are, however, rare.

(iv) *Turreted*: This rare form of *scallop* was first described by Allen (1971a: Figs. 83C–E therein), who generated it experimentally, but it has never been described in the field. In studies undertaken for this report, it was identified at Sleightholme Beck. *Turreted scallops* arise through an "instability to transverse disturbances" within new *scallops* that have been created at the streamwise crest between two preexisting *scallops* lying side by side (Allen, 1971a). The result of this instability is the formation of a series of closely spaced transverse crests within the new *scallop*, giving it a turreted appearance (Fig. 133).

4.3 Other Solutional Forms

(a) *Vertical ridges.* These are rounded ridges only a few millimeters across and a few millimeters in relief that form on near-vertical faces below horizontal ledges (Fig. 134). They typically cover large areas of each face. It seems likely that they develop in response to a slow but steady supply of water draining from the ledge, distributed across the vertical face rather than supplied at a point, and flowing down the face as a film. They are similar to the "decantation flutings" described by Ford and Williams (1991: p. 388) but are less well developed and on a smaller scale. *Vertical ridges* can be superimposed upon *solution ripples*.

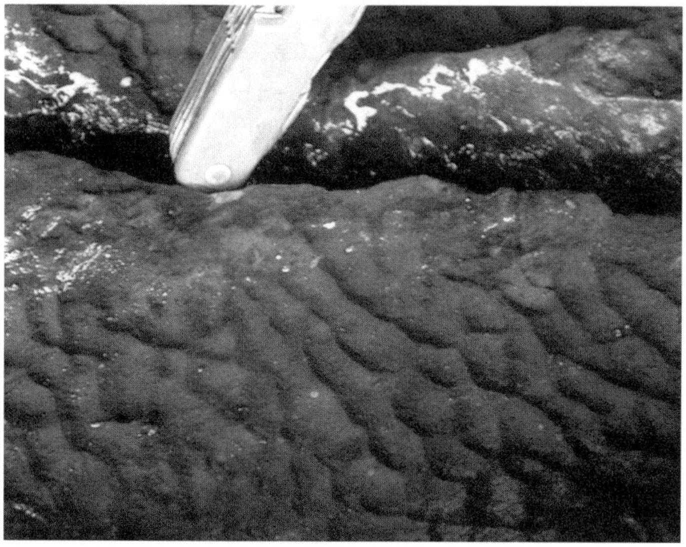

Figure 132. *Two-dimensional scallops* (below and to right of penknife). Sleightholme Beck, UK. Flow from bottom left to top right. Limestone.

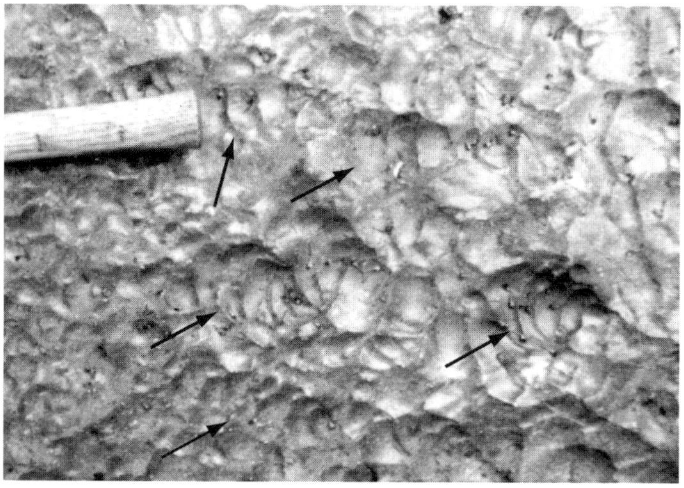

Figure 133. *Turreted scallops* (arrowed). Sleightholme Beck, UK. Flow from left to right. Limestone. The marks on the hammer shaft are 5 cm apart.

Figure 134. *Vertical ridges*. Ease Gill, UK. Limestone. Penknife for scale.

(b) *Solution ripples.* These are transverse ripple-like features that form on near-vertical to overhanging surfaces (Fig. 135) over which water flows as a film or thin sheet. They have a wavelength of up to 3 cm (Allen, 1982: p. 247). They are morphologically similar to *two-dimensional scallops*, on which basis Allen proposes a relationship between them, but they are responding to entirely different flow conditions. The *solution ripples* are orientated transverse to the flow of water on the surface on which they occur, and therefore the crests are not necessarily horizontal. If they form under spray water flowing passively down the steep face under gravity as a film, the crests will be horizontal. However, they also occur on faces subject to the oblique impact of jets of water, in which case the crests can be at any angle to the horizontal, depending on the orientation of the jet. Both types of *solution ripples* are present at Sleightholme Beck.

(c) *Decantation runnels.* This is a term used to describe small channels carrying a steady supply of water released from a point source (Ford and Williams, 1991: p. 376, 386). They are common in the channel margins, where vegetation and soil cover can form point sources of water. They may be straight or meandering (maänderkarren) and typically connect several *concave features* to form a continuous *runnel* (Fig. 136). In some cases, they are widest at the source of water and narrow with distance from it.

(d) *Longitudinal ridges.* These are assumed to be a *solutional form* because they were observed only at two sites, both of which

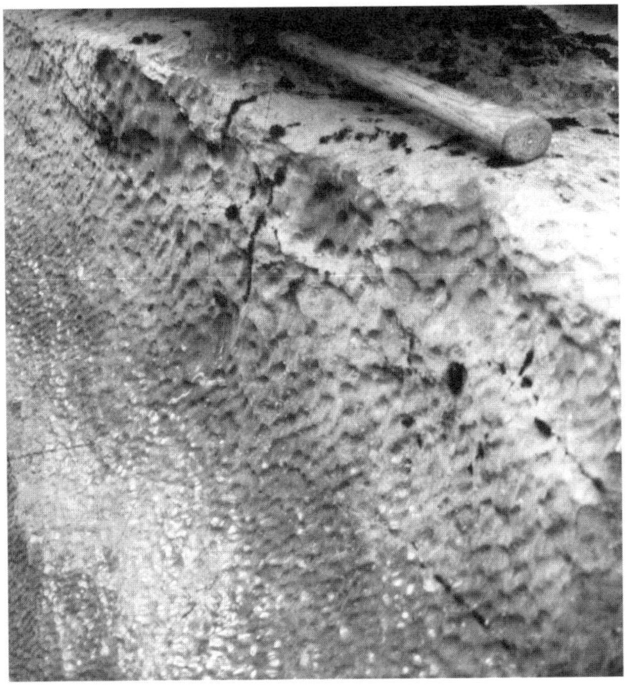

Figure 135. *Solution ripples*. These bedforms are on a near-vertical face that experiences sheet flow from free-falling jets of water directed at the face. The radial pattern of crest lines indicates the divergent nature of the directions of the jets that impact upon the surface. Sleightholme Beck, UK. Flow from top left to bottom right. Limestone. The marks on the hammer shaft are 5 cm apart.

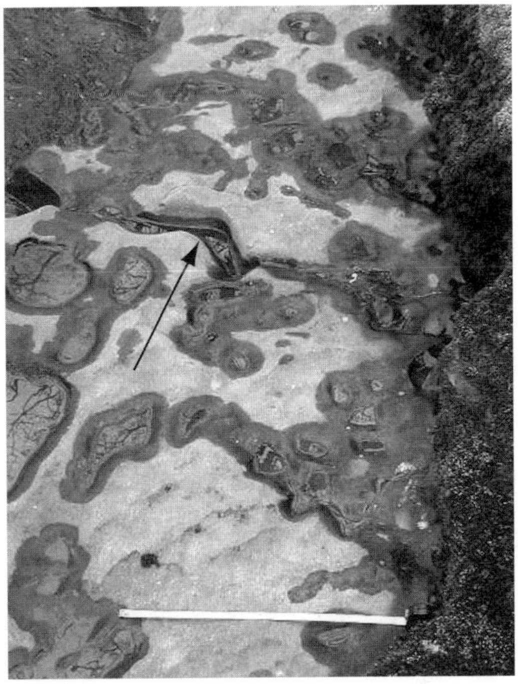

Figure 136. *Decantation runnel* (arrowed). This *runnel* is fed from a point source in the moss-covered area on the right and drains into the main part of the channel on the left. River Dee, UK. Flow in the main channel is from top to bottom. Limestone. The tape measure is extended to 50 cm.

are in limestone. They occur on both vertical and horizontal faces. The *longitudinal ridges* are parallel to flow, although their orientation is sometimes influenced by the rock structure. *Longitudinal ridges* are spaced 2–10 cm apart and are continuous typically for up to 30 cm on the channel floor, but may be longer on vertical walls. They have a relief of up to 2 cm. They exist in two subtly different forms: (1) as simple *longitudinal ridges*, either on surfaces devoid of other bedforms or within a field of *scallops* (but unrelated to the *scallops*) (Fig. 137A), and (2) as *longitudinal ridges* intimately associated with *scallops*, in which case they are formed through an in-phase arrangement of the *scallops* (Fig. 137B). In the latter case, the *longitudinal ridges* arise through a kind of constructive interference between the *scallop* crests, in which longitudinally orientated crest sections become elevated. On vertical faces adjacent to *steps* in the channel floor, *longitudinal ridges* are curved and dip downwards at a high angle, following the curving flow pathlines over the step (Fig. 137A). Both sites containing *longitudinal ridges* are in small gorges where gradient and flow velocities are high, and this is clearly a factor in the development of *longitudinal ridges*. They may be analogous to the sand ridges with a transverse spacing on the decimeter scale observed by Karcz (1973), which were found to replace ripples at high current velocities. It is also possible that they are related to the streaky structure observed within a fully rough boundary layer (Grass et al., 1991; Defina, 1996; Grass and Mansour-Tehrani, 1996).

(e) *Microrills*. A well-known karren form, *microrills* are small-scale, densely packed solutional *furrows*, typically 1 mm across. They range from sinuous and anabranching to straight, depending on the slope (Ford and Williams, 1991: p. 378). They are found in some limestone bedrock channels on surfaces that experience significant subaerial weathering, and that are protected from intense fluvial erosion when submerged, such as the lee slopes of *SCHF* (*sharp-crested hummocky forms*) (Fig. 138). It must be assumed that strong fluvial erosion on other surfaces prevents *microrills* from forming.

5. Tool Marks

This small group of bedforms is the second to be based on inferred mechanisms. *Tool marks* are illustrated schematically in Figure 139.

5.1 Percussion Marks

Percussion marks are small depressions (up to a few centimeters), each created by the chipping away of a rock fragment from a boulder or rock projection by the impact of a large bedload clast.

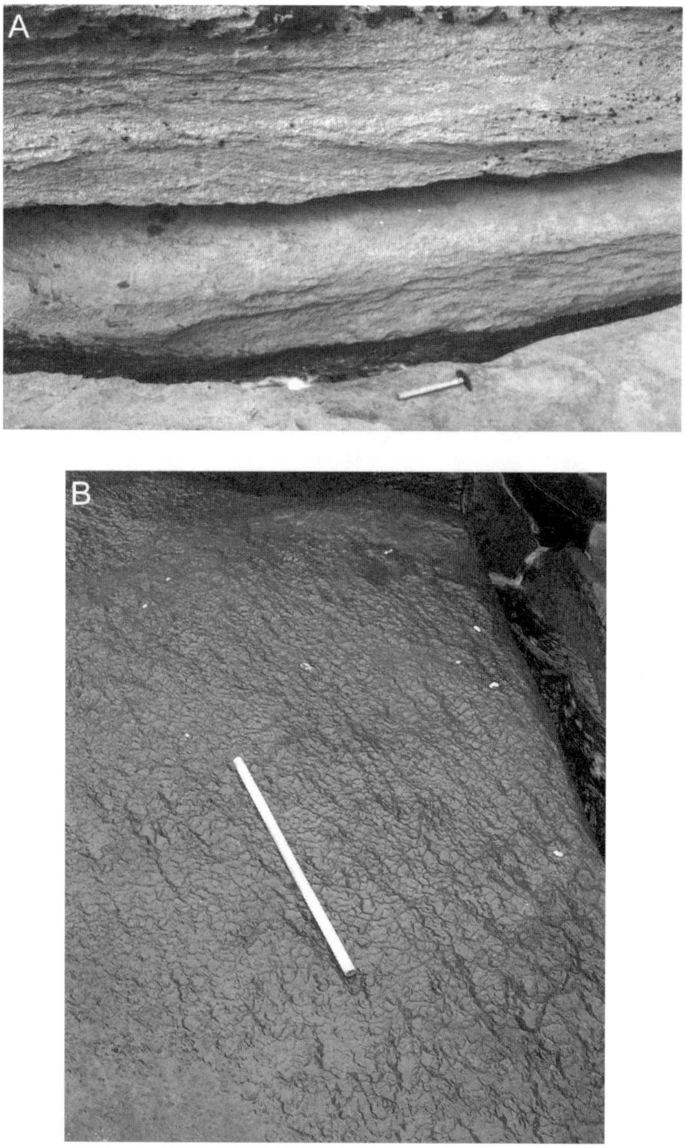

Figure 137. A: *Longitudinal ridges* on a vertical face devoid of other bedforms. The downward-dipping ridges in the center and center left of the picture reflect the flow pathlines of water over the steeply sloping channel floor to the left of the hammer. Sleightholme Beck, UK. Flow from left to right. Limestone. The hammer is 30 cm long. B: *Longitudinal ridges* in horizontal surface containing *scallops*. Sleightholme Beck, UK. Flow from bottom right to top left. Limestone. The scale is 60 cm long.

(a) *Edge percussion marks.* In such marks, the rock fragment is chipped away from the angular edge of a boulder or projection, leaving an indentation with sharp margins (Fig. 140A). The margins of the *percussion mark* may be irregular or straight. If there is a rock structure, such as cleavage, the margins will tend to follow this.

(b) *Face percussion marks.* In this case, the rock fragment is chipped away from the face of a boulder or projection, leaving a circular or slightly ovoid shallow depression usually no more

Figure 138. *Microrills* on the lee face of a *sharp-crested hummocky form*. Nahal Zin, Israel. Flow over the *hummocky form* is from top to bottom. Limestone. Coin for scale.

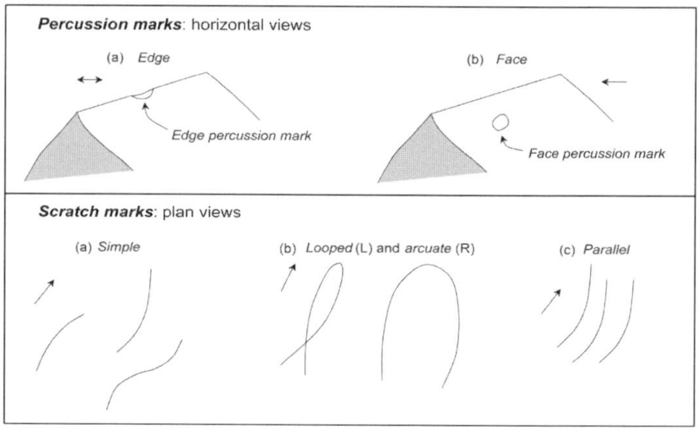

Figure 139. *Tool marks*. Arrows indicate flow directions.

than a centimeter or two in diameter but occasionally larger (Fig. 140B). Newer examples have slightly irregular margins, but older examples appear to exhibit more rounded rims and occasionally seem to become loci for enhanced erosion, as they can be deeper than any impact fracturing could induce directly. In older examples, the circular or ovoid planform may also become modified by erosion focused on the downstream lip, and the *percussion mark* may become more *flute*-like.

(c) *Cavitation percussion marks.* Although cavitation requires very high fluid velocities, many authors support the idea that it is an important process in extreme events (e.g., Baker, 1973; Sato et al., 1987; Wohl, 1993; Shaw, 1996; Baker and Kale, 1998; Hancock et al., 1998; Tinkler and Wohl, 1998a), and possibly even in annual floods in very large rivers (Gupta et. al., 1999; Whipple et al., 2000a). Shaw (1996) described *percussion*

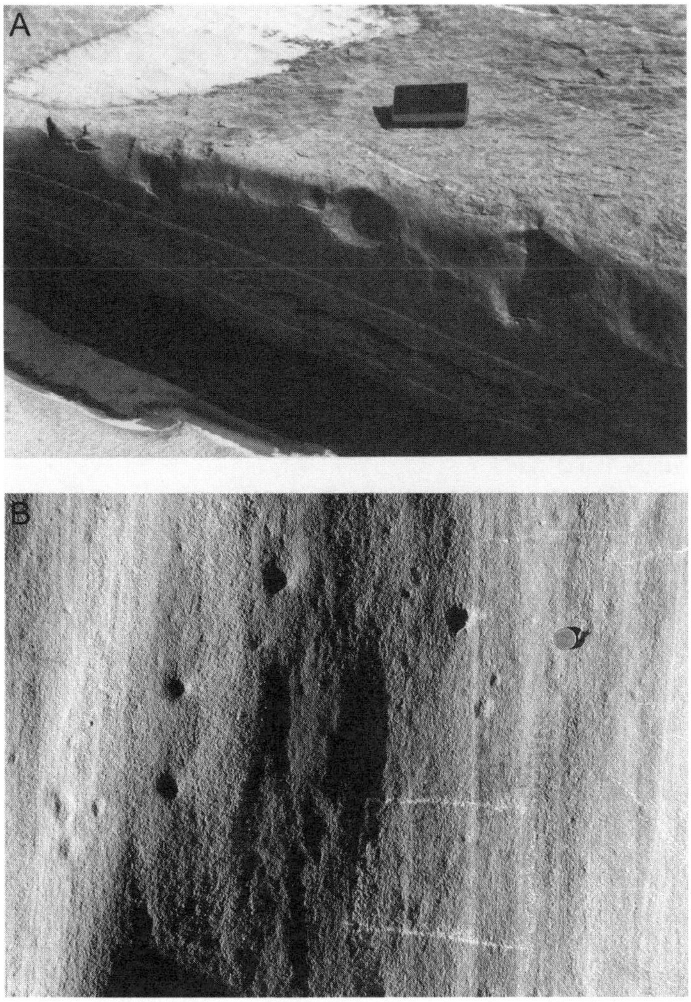

Figure 140. A: *Edge percussion marks*. Nam Mae Khlang, Thailand. The vertical face looks upstream. Gneiss. The notebook is 15 cm long. B: *Face percussion marks*. The marks immediately below and to the left of the coin are relatively fresh, while the larger marks with dark shadows are older, deeper and more rounded due to enhancement by abrasion. Nahal Shani, Israel. Fine-grained sandstone.

marks in quartzite that he considered to be due to cavitation in subglacial meltwater floods in which fluid reached velocities of ~60 ms^{-1}. These marks occur in two forms: as cones exhibiting concentric fractures (Fig. 141A), and as small impact marks that often occur in rings 5–10 cm across (Fig. 141B). Such rings are interpreted as the footprint of cavitation within the cores of small vortices.

5.2 Scratch Marks

Scratch marks are most often found in *plane-bed* channels with sculpted forms, where coarse bedload is not restricted to the thalweg, and where clasts in transit strike against the stoss slopes of bedforms.

(a) *Simple scratch marks.* These are linear (straight or slightly curved) or irregular, isolated *scratch marks*. They may occur in high densities on a surface, but each is unrelated to any other mark.

(b) *Looped* and *arcuate scratch marks.* These forms are common on the upstream-facing slopes of depressions, where clasts residing temporarily within the depression are dragged one or more times up that face but, failing to reach the rim, slide down again under their own weight, generating either a *looped* or an *arcuate scratch mark* (Fig. 142).

(c) *Parallel scratch marks.* A clast with two or more points in contact with the surface will generate multiple, parallel *scratch marks*. The marks may be *linear*, *arcuate*, or *looped* (Fig. 142).

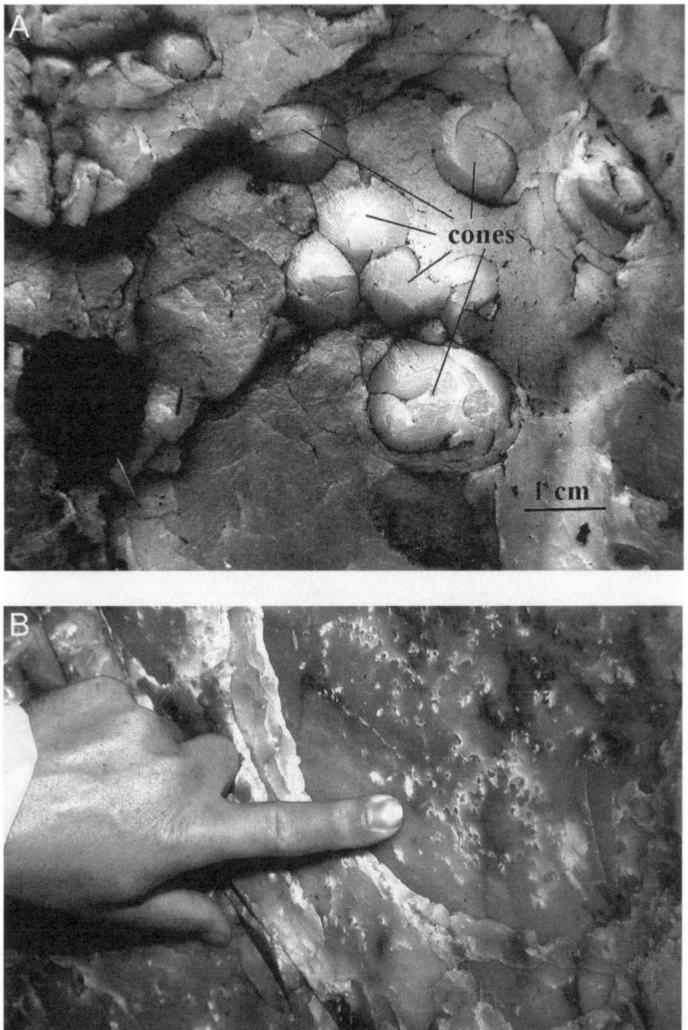

Figure 141. A: *Cavitation percussion marks* in the form of cones formed by concentric fractures. Colline Blanche, Canada. Quartzite. Photo courtesy of John Shaw. B: *Cavitation percussion marks* in the form of rings of impact marks. Colline Blanche, Canada. Quartzite. Photo courtesy of John Shaw.

Figure 142. *Parallel scratch marks*. Most of the *scratch marks* are *linear*, but some in the upper part of the photograph are *arcuate*. River Dee, UK. Flow from bottom left to top right. Limestone. The hammer is 30 cm long.

6. Large-Scale Sculpted Features

This typology has so far dealt with small-scale sculpted forms, whose characteristic dimension is much less than the channel width. The present section, however, briefly covers bedforms whose size is comparable with the channel width, or which ranges up to this size. Following the approach of Wohl (1998), large-scale features are subdivided into those which result in uniform and variable bed gradients respectively.

6.1 Uniform Bed Gradient

(a) ***Discordant plane bed.*** Erosion is approximately uniform across the channel. The channel floor is planar in a macroscopic sense, with a uniform gradient, a regular rectangular to trapezoidal cross section (Wohl, 1998), and little or no structural control at this scale (Figs. 2C and 2D). *Discordant plane beds* therefore represent a large-scale sculpted form. In detail, however, the bed may show a structurally controlled morphology of considerable relief (Fig. 1A), or well-developed sculpted forms with some degree of structural influence (Fig. 157). *Discordant plane-bed* channels are described by Suzuki and Ikeda (1994), Itakura and Ikeda (1997), and Wohl and Ikeda (1998). See also the terminology section of this report.

(b) ***Inner channel*** **and** ***bedrock benches.*** Incision of the thalweg into the floor of the main channel produces a steep-sided *inner channel* flanked by raised *bedrock benches* (Fig. 143). This is indicative of erosion that is highly nonuniform over the cross section of the channel (Wohl, 1998). These features are described by Bretz (1924), Baker (1973), Nemec et al. (1982), Baker and Pickup (1987), Kale and Shingade (1987), Wohl (1992a, 1992b, 1993), Wohl et al. (1994), Baker and Kale (1998), Tinkler and Wohl (1998a), and Wohl and Achyuthan (2002). In the experience of the authors, *inner channels* normally appear to be formed by the migration of *knickpoints* (see section 6.2a), with a *knickpoint* occurring at the head of each *inner channel*, suggesting that both of these features are part of a single phenomenon. *Knickpoint* migration is often effected through the coalescence of *potholes*, in which case the resulting *inner channel* has complex, sinuous walls with projecting vertical ridges (Elston, 1917, 1918; Jennings, 1983; Wohl, 1993, 1999; Kunert and Coniglio, 2002) (Fig. 22B). Sometimes the wall sinuosity is regular and repetitive (see section c below). However, Shepherd and Schumm (1974) observed experimentally *inner channels* forming through the coalescence of *groups of parallel-sided furrows*. Wohl and Ikeda (1997) observed in flume experiments the development of *parallel-sided furrows*, a broad shallow channel with undulating walls, and finally a deep and more undulating *inner channel* as gradient increased.

(c) ***Inner channel*** **and** ***slot canyon with out-of-phase undulating walls.*** These consist of *inner channels* with regular wall sinuosity that is out of phase (but not necessarily in antiphase), creating repetitive variations in channel width (Wohl, 1999; Wohl et al., 1999; Wohl and Achyuthan, 2002). Examples with extremely small width:depth ratios are commonly termed *slot canyons*; in such channels, the walls make up a larger proportion of the channel boundary than does the floor, and the "wallforms" are hypothesized to regulate flow energy expenditure in a manner analogous to bedforms on the floor of a wide channel (Wohl, 1998, 1999; Wohl et al., 1999). *Inner channels with out-of-phase wall undulations* commonly occur downstream of *knickpoints* or *knickzones* in homogeneous substrates (Wohl et al., 1999). Such *knickzones* are characterized by the presence of sinuous *runnels*

Figure 143. *Inner channel* and *bedrock benches*. Nahal Shani, Israel, looking upstream. Fine-grained sandstone. Person for scale.

containing offset *lateral potholes* (termed *runnels with alternating scour* in this typology, section 1.2f). *Knickzone* migration is effected through the enlargement of the *furrow* and growth and coalescence of the *potholes*, which process also produces the *wall undulations*. The *wall undulations* are preserved in the *inner channel* after the *knickzone* has migrated upstream (Wohl, 1998). An *inner channel with out-of-phase wall undulations* was noted by Wohl (1993) as the culmination of a downstream sequence of forms that began with a single *furrow with regularly spaced depressions* (see sections 1.2c and 1.2f). This *furrow* and its depressions grew with distance downstream, becoming a *runnel with alternating scour*, and ultimately enlarging into an *inner channel with wall undulations*.

(d) *Inner channel with in-phase undulating walls.* In some *inner channels*, it has been noticed by the authors that repetitive *in-phase wall undulations* occur (Fig. 144). In common with the out-of-phase variety, these *wall undulations* arise through the enlargement and coalescence of *potholes* in the channel floor in the region of accelerating flow immediately above a *knickpoint*, and also represent a mechanism of *knickpoint* retreat and *inner channel* formation. However, in the in-phase case, the *potholes* develop in an in-line series, and each *pothole* grows until it extends across the full width of the *inner channel*, thus creating undulations in each *inner channel* wall that are in phase with those in the opposite wall. A *downstream-facing step* is formed in the floor of the *inner channel* where one *pothole* passes into the next one downstream. An interesting feature of these *potholes* is that, once they extend across the full width of the *inner channel*, they must contain two counterrotating vortices in the two halves of the depression separated by a central region of strongly advecting flow, rather than the single vortex assumed to exist in other *simple potholes*.

Figure 144. *Inner channel with in-phase undulating walls*. Individual *potholes* extend right across the *inner channel*, forming in-phase undulations in its sidewalls. Ease Gill, UK, looking downstream. Limestone. The large *pothole* in the foreground is ~1.5 m across.

(e) *Yardang.* Well known in aeolian environments (e.g., Whitney, 1983; Summerfield, 1991: p. 240–241; Livingstone and Warren, 1996: p. 32–34), *yardangs* also exist in aqueous environments. Typically, they are ridges parallel to the flow with a blunt nose and a tapering tail, and with the widest and highest point toward the upstream end, although in aeolian environments their morphology is variable (Livingstone and Warren, 1996: p. 33). They may be up to several tens of meters in length. They may also be undercut at the base, especially in strongly bedded rocks consisting of alternating beds of contrasting resistance (Fig. 145). In aeolian environments, this is thought to be the result of enhanced erosion in the region of saltating grains close to the ground (Summerfield, 1991: p. 240); in rivers, it may indicate that erosion by bedload is important. Bryant (2001: p. 87) described *yardangs* (which he referred to as "flutes") from coastal environments due to tsunamis.

6.2 Variable Bed Gradient

(a) *Knickpoint* and *knickzone*. *Knickpoints* are upstream-migrating points of enhanced energy dissipation and erosion. They may take the form of vertical or even undercut waterfalls, or of short steep reaches of channel referred to as *knickzones* (Wohl, 1998). In plan view, *knickpoints* may run straight across the channel or take the form of V-shaped headcuts. They are indicative of active local incision, which may have been initiated by a relative fall in base level (Righter, 1997; Knighton, 1998: p. 244) and/or a decrease in sediment supply (Wohl et al., 1994).

Figure 145. *Yardang*. This bedform is in a tidal reach of the River Severn, UK, in an ebb-dominated (current from left to right) area. Mudstone. The scale is 60 cm long.

In layered rocks, they may also occur at outcrops of resistant strata, which provide local base-level control (Foley, 1980a; Wohl et al., 1994; Knighton, 1998: p. 244), and at *steps* that arise as a result of the relationship between channel gradient and the dip of the strata (Miller, 1991). *Knickpoint* and *knickzone* migration is an important mechanism in the formation of *inner channels* (Baker, 1973; Wohl, 1993; Wohl et al., 1994) and of erosional escarpment retreat (Weissel and Seidl, 1998). The substrate exerts a strong control on whether near-vertical *knickpoints* can be maintained as equilibrium forms during headward migration. In homogeneous rocks, *knickpoints* tend to erode to a lower angle and become indistinct as they migrate (Gardner, 1983; Wohl, 1998). Similarly, thinly bedded rocks and cases where bedding strikes parallel to the channel do not facilitate the formation of *knickpoints*. However, in strongly layered or jointed rocks that are appropriately orientated relative to the channel, they may maintain their steep form until longitudinal profile adjustments to base-level lowering (or other change) can be made (Wohl, 1993, 1998; Wohl and Ikeda, 1998).

(b) *Cataract.* A stair-like succession of *knickpoints* (Wohl, 1998).

(c) *Step.* *Step* faces are vertical or near-vertical, and water flowing over them does not remain in contact with the face. They are often structurally controlled, for example, by bedding, and may be *forward-* or *backward-facing*, depending on the relative slope of the strata and channel (Miller, 1991; Wende, 1999). However, it is also thought that *steps* may be formed in homogeneous substrates (see [j] below), either isolated or as part of a *step-pool* sequence, as a result of the nonlinear erosive response of the bed (Parker and Izumi, 2000).

(d) *Chute.* A *chute* is generally accepted as a smooth, steep surface at an angle less than vertical that provides no effective restraint to the flow, but over which the flow remains substantially in contact with the rock surface. At the point of release at the top of such a surface, the flow is critical (Henderson, 1966: p. 29–57), and it is therefore supercritical over the length of the *chute*. Flow down a *chute* is often termed "shooting flow." The *chute* may terminate in a pool, which contains a hydraulic jump and possibly a *gouge hole* (see below), or in a free-falling *step* or waterfall. *Chutes* are variable in form and range from convex to concave surfaces in both longitudinal and transverse sections. They may occupy part or all of the channel width. *Chutes* occur in depositional sedimentary environments (Leeder, 1982: p. 92; Fralick, 1999), incising bedrock channels (Wohl and Ikeda, 1998), and as s-forms (Glasser and Nicholson, 1998).

(e) *Bifurcating chute*. It is often found that broad surfaces forming *chutes* become divided at their lower end by a rounded ridge that focuses water to either side of it (Fig. 146). It appears that laterally uniform supercritical sheet flow is therefore unstable and will tend to become channelized. Izumi and Parker (2000) applied linear stability analysis to sinusoidal lateral perturbations in a convex slope and found that they were unstable within a range of wave numbers of the perturbation, leading to incipient channelization, with a characteristic spacing of ~1000 times the critical flow depth. This spacing is much greater than that of the linear depressions within a *bifurcating chute*, although the analysis applies to subcritical rather than supercritical flow.

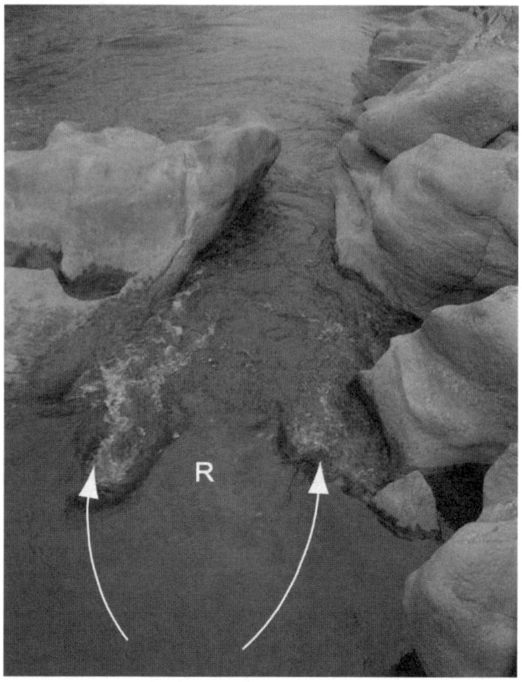

Figure 146. *Bifurcating chute*. Note the central ridge (R) in the *chute*, which diverts flow to either side (arrows). Nam Mae Chaem (Ob Luang), Thailand, looking downstream. Granitic gneiss. The *chute* is ~3 m wide.

Figure 147. Several *gouge holes* below a gently sloping *chute*. Two large *gouge holes* can be seen near the center of the picture, while three smaller, elongated ones are in the center right of the picture. Birk Beck, UK. Flow from top left to bottom right. Medium-grained sandstone. The scale bar is 30 cm long.

(f) *Plunge pool* and *gouge hole*. *Knickpoints*, *steps*, and *chutes* often have at their foot a depression eroded by the high-velocity water flowing over them. Alexander (1932) reserved the term *plunge pool* for depressions associated with the impact of a water jet upon a rock surface orientated normally to it, while he termed hollows generated by the oblique incidence of water upon the rock (such as water exiting a *chute*) *gouge holes* (Fig. 147). Alexander (1932) classified both as types of *potholes*, which is regarded as erroneous in this report, and his postulated flow structures for *plunge pools* and *gouge holes* should be viewed with skepticism. However, there is merit in distinguishing between scour holes formed by water jets of different orientation, and the terms *plunge pool* and *gouge hole* are adopted here with the same meanings as those applied by Alexander (1932). *Plunge pools* are associated with *steps* and approximately vertically falling water. The momentum of the incident water, which is orientated close to the vertical, is almost entirely dissipated in the *plunge pool*, which is characterized by extremely turbulent and chaotic flow and highly aerated water. However, in *gouge holes*, which are shallower than *plunge pools*, and which are associated with *chutes*, the water enters at the upstream end at an oblique angle to the horizontal and leaves at the downstream end, retaining a considerable portion of its downstream momentum. *Gouge holes* are elongated in the direction of flow and typically contain a hydraulic jump (although this would be drowned out at high stage). There is clearly a continuum between these two bedforms in terms of the angle of incidence of the water jet and the flow structure, and associated with this a continuity of form, but there is no obvious continuum in either sense between these bedforms and *potholes* as Alexander (1932) suggests there is. Jacob et al. (1999) describe *plunge pools* from the paleo-Orange River, South Africa.

(g) *Step-pool system*. This type of channel topography is assumed here to be synonymous with "chute-and-pool." Wohl and Ikeda (1997) make a distinction between these terms in their experiments on the erosion of a cohesive substrate, obtaining true, near-vertical, abrupt bed *steps* only when the substrate possessed a pronounced structure. However, in nature, *steps* are often roughly hewn, imperfect features over which water does not remain continuously in free fall. Furthermore, *chutes* often possess a near-vertical surface with which the water is no longer in contact at their lower end, and *steps* often possess a *chute*-like surface in their upper part. The important feature of a *step-pool* system is a repeating and pronounced variation in channel bed gradient that results in alternating subcritical and supercritical flow, with associated hydraulic jumps. Therefore, any distinction between *step-pool* and chute-and-pool systems would appear to be artificial and trivial. Bedrock channel morphology has traditionally been regarded as controlled by the rock structure (Ashley et al., 1988). That this can control the distribution of *steps* in *step-pool* systems is evident (Miller, 1991; Duckson and Duckson, 1995; Wohl and Ikeda, 1997; Wohl, 1998). However, Kodama and Ikeda (1984), Wohl and Ikeda (1997), and Koyama and Ikeda (1998) have produced repeating *steps* experimentally in homogeneous cohesive substrates, while Wohl and Grodek (1994), Koyama and Ikeda (1998), and Wohl and Ikeda (1998) describe field examples of *step-pools* in homogeneous substrate. The present authors have also observed *step-pools* in homogeneous granite (Fig. 148). Parker and Izumi (2000) generate cyclic *steps* in numerical simulations of the erosion of a cohesive homogeneous substrate. Therefore, perhaps the terms *structurally* and *hydrodynamically controlled step-pool systems* might be applied to the two types of cases. Wohl (2000) demonstrates the importance of the ratio of substrate resisting forces to hydraulic driving forces in determining the ratio of *step* height to *step* length. *Step-pool* morphologies can also occur on a smaller scale within *runnels* (section 3.2b).

(h) *Pool-riffle/pool-rapid system*. Quasi-periodic variations in bed gradient comprising *pool-riffle* sequences are well known from alluvial rivers, but directly analogous features may also be found in bedrock channels and mixed bedrock/alluvial rivers (Keller and Melhorn, 1978; Baker and Pickup, 1987; Wohl, 1992a). These are comparable to their alluvial counterparts morphologically and are also indistinguishable in terms of their spacing, which averages between five and seven channel widths (Keller and Melhorn, 1978).

(i) *Breached pools*. There are two varieties of *breached pools*. *Pools* within a *step-pool* system may become breached by a headcut progressing from the immediate downstream

Figure 148. *Step-pool* system. Nishi-zawa, Japan, looking upstream. Granite. The channel is ~6 m wide.

Figure 149. A: *Breached pool*. A headcut has eroded from the lower pool (in which the photographer is standing) into the pool above (immediately behind the standing figure), through the intervening step (S). Than Rattana, Thailand, looking upstream. Andesite. B: *Breached pool*. The pool in the center of the photo has been undercut and breached by the pool below (lower left of photograph). The pool in the center is now completely drained, and the lip between them is left as a *natural arch*. Wadi Hazazon, Israel. Limestone. The upper pool is ~6 m wide; the vertical drop to the base of the pool is ~20 m.

pool through the intervening *step* (Fig. 149A). Although the morphology of the *pool* remains essentially unaltered, low stage drainage is enhanced, such that the low stage volume of the *pool* is reduced. More rarely, the base of *pool* may be breached, although the *step* on the downstream side remains intact (Fig. 149B). This occurs when the *pool* immediately downstream becomes excessively deep, and the entry flow is nearly vertical. In such cases, in addition to a strong upwelling current progressing downstream, there is a strong countercurrent acting against the headwall of the downstream *pool*. In time, this can cause the recession of the base of the headwall until a breakthrough is made into the base of the upstream *pool*. In this case, the upstream *pool* is drained completely. This process may be the precursor to the formation of a channel-scale *natural arch*.

(j) *Convex surface*. Some channels in relatively homogeneous substrates that lack closely spaced joints exhibit groups of convex-up, gently rounded surfaces (Fig. 150). These surfaces may be on the scale of the channel width, or be a fraction of it, but they are at least an order of magnitude larger than typical *hummocky forms*. *Convex surfaces* are illustrated (but not described) by Kasai et al. (2004: Fig. 2B therein).

(k) *Incised convex surface*. In some cases, *convex surfaces* may be found with *runnels* or *inner channels* incised into them, although only two such examples were seen by the authors. They are convex in both cross and longitudinal sections, with a smaller radius of curvature in the transverse direction, forming a broad, sloping humpbacked ridge parallel to flow (Fig. 151). These surfaces would both, at high stage, have formed *chutes*, but at the low stage observed, flow was restricted within the *runnel* or *inner channel* incised into them. The most interesting point about these features, however, was that the *runnel* or *inner channel* flowed down the apex of the ridge, rather than taking the steepest route down the side of the *convex surface*. In one example, the *runnel* had *cusped margins* (section 1.2f) at its upper end.

Figure 150. *Convex surfaces*. Huai Nang Rong, Thailand, looking downstream. Rhyolitic agglomerate. Person for scale.

Figure 151. *Incised convex surface*. Tributary to Sichiri-gawa, Japan, looking upstream. Fine-grained sandstone. Person for scale.

FURTHER OBSERVATIONS AND DISCUSSION

Principles Applying to the Morphology of Bedrock Bedforms

The preceding sections have described all the bedforms observed by or known to the authors that, in spite of the enormous variety of bedrock bedforms, can be considered as distinct bedform types owing to either their frequency of occurrence or their difference in key morphological aspects from other, similar bedforms. However, in the observations of bedrock channels made for this study, it was found that erosional forms illustrate more than simply the wide variety of forms that exist. In fact, four general principles can be said to apply to the development of bedrock bedforms, one of which has been highlighted to some extent already. These principles will be referred to as continuity of form, convergence of form, the constructive interference of the flow structures of contiguous bedforms, and the prevalence of sharp transverse crests.

1. Continuity of Form

Continuity of form is the phenomenon of the gradation between end-member bedforms, producing a spectrum of closely related features. Such a spectrum is produced as one or more morphological parameters of the features varies from one extreme to the other in a continuous manner. The end members are often far apart, which is why this is an important phenomenon. Continuity of form arises through the complex and nonuniform flow present within bedrock channels, which results in the existence of a wide range of flow patterns and hydrodynamic conditions. This in turn results in the singular nature of bedrock bedforms. In some cases, continuity of form may also arise through the evolution of one form into another with time. These phenomena create almost infinite subtlety in the variation among bedforms, allowing distinct forms to grade continuously into each other via many small intermediate steps. Continuity does not negate the purpose of this study because it is important to document the existence of the variety, and to describe its limits and logical reference points within it. Continuity of form is operative in all classes of bedforms. Some examples and consequences of the phenomenon are described below. It is important to note, however, that continuity between bedforms in space, even if it occurs within a short distance, may or may not imply a sequence of development from one form to another with time. Continuity of form has been observed in s-forms by Glasser and Nicholson (1998) and Kor and Cowell (1998).

(i) The single most important example of continuity is the gradation between all types of *concave features*. The spectrum of length:width ratios leads continuously from *circular potholes* to highly elongated *parallel-sided furrows*. The spectrum of degrees of erosion of the downstream rim leads continuously from *short furrows* to *flutes*. A spectrum of length:width ratios also appears to form a continuum between *flutes* and *parallel-sided furrows*, which are often generated by some defect and which are *flute*-like at their upstream ends.

(ii) *Scallops* illustrate various continua of form. These exist between *two-* and *three-dimensional scallops*, between *directional* and *nondirectional scallops*, and between *space-filling solution pits* and *three-dimensional scallops*.

(iii) There is continuity of form between the various types of *solution pits*. In particular, *shallow circular*, *elongated*, *open*, and *extended open solution pits* form a logical sequence of increasing length:width ratio and amount of erosion focused on the downstream part of the rim.

(iv) *SCHF* (*sharp-crested hummocky forms*) are gradational with *scallops*. This is discussed further below.

(v) *Reversed furrows* have variable slopes along their axes. As they tend toward the vertical, they approach the state of a *lateral pothole*. A similar change may be effected by means of the *reversed furrow* becoming wider, more strongly tapered or shorter and deeper, so that it is no longer a linear feature in plan view. *Oblique sloping furrows* grade into *linear lateral potholes* as the slope of the surface on which they occur, and therefore the slope along the *furrow* axis, increases toward the vertical.

(vi) *Potholes* grade into *lateral potholes* as the slope of the surface in which they occur increases from the horizontal. Within the class of *lateral potholes*, there is a spectrum of forms between *equant* and *linear* types, and between sharp- and round-rimmed types. Furthermore, as the axis of a *linear lateral pothole* tends toward the horizontal, it approaches the state of a *transverse furrow*. Indeed, individual bedforms may be *transverse furrows* in their upper reaches and lead downwards into a *lateral pothole*.

(vii) *Sharp-nosed obstacle marks* are gradational with *obstacles with secondary sculpting*. This occurs by means of the ridge or "sharp nose" of the *obstacle* becoming increasingly skewed to one side. In some examples, it is no longer at the front pointing forward but is shifted around to the side of the *obstacle*. In this position, the secondary sculpted form on the downstream side of the ridge begins to erode into the side of the *obstacle*, and the bedform therefore becomes an *obstacle with secondary sculpting*.

(viii) *Irregular sculpting* on the lee faces of projections grades into a more organized style of sculpting consisting of *short furrows* as the orientation of the surface rotates out of the lee of the projection and becomes instead parallel to the flow. When *irregular sculpting* is found close to the corner of square-cross-section channels, a similar change in its morphology occurs with increasing distance from the corner.

2. Convergence of Form

Convergence, also known as equifinality, is defined as different causes and processes having similar end results (Schumm, 1991: p. 58). The idea of convergence of form in bedrock channels is based on the fact that a small number of "preferred" morphologies occur repeatedly in quite different circumstances. This results in bedforms that, although they have arisen through different evolutionary paths and are not identical, are yet analogous or contain within them analogous structures. Having reached this common state, it is assumed that their subsequent behavior and associated flow structures will be similar. The convergence of form in bedrock channels can be seen to occur in at least three slightly different ways, described conceptually below (see Fig. 152). In the following descriptions, B_1 and B_2 are bedforms where B_1 either is analogous to B_2 or contains within it a structure that is analogous to B_2.

(i) Route 1: Starting from either a featureless or *irregular surface* with no bedforms (NB), different bedforms begin to develop along paths P_1 and P_2. Although these paths initially result in quite distinct forms, after some time, they result in convergence, and the bedforms become similar (B_1, B_2).

(ii) Route 2: Again, starting from a surface with no bedforms (NB), different bedforms begin to develop along paths P_1 and P_2. Path P_1 leads initially to bedform B_0 and subsequently, with continued development along the same path, to a new bedform B_1. Path P_2, however, leads directly to the development of bedform B_2. Thus the paths P_1 and P_2 are again convergent, but one path converges with the other via the intermediate state B_0.

(iii) Route 3: Starting from NB, different bedforms begin to develop along paths P_1 and P_2. P_1 again leads to an initial bedform B_0. However, continued development along the same path now leads to a mature, well-developed example of the bedform B_{0M}. On reaching this state, a threshold is crossed whereby the bedform substantially changes the local flow pattern and consequently begins to develop along a new path P_1', which transforms B_{0M} into the new bedform B_1. Path P_2 again leads directly to bedform B_2. Paths P_1 and P_1' are therefore convergent, via intermediate states, with P_2, but in this case the bedform itself (B_{0M}) plays an active role and triggers the convergence.

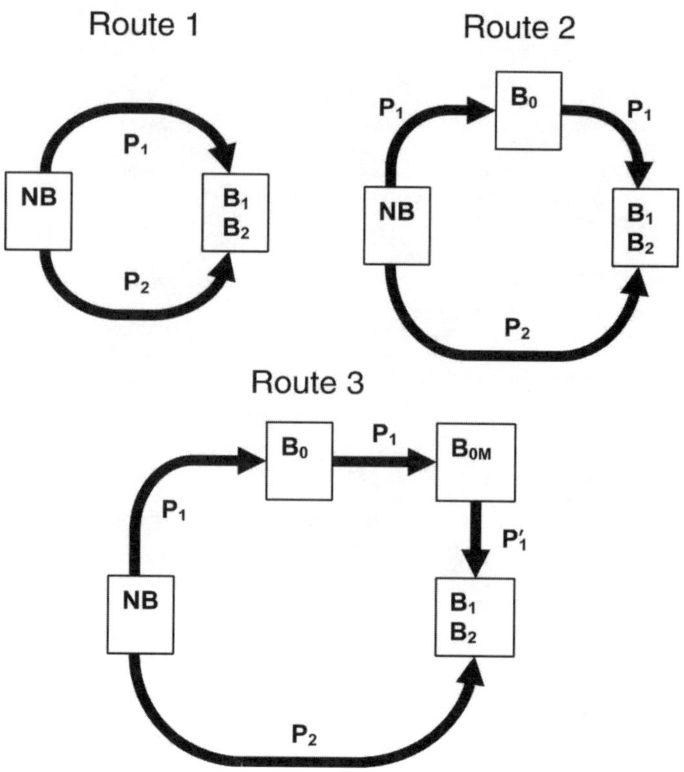

Figure 152. Schematic diagrams illustrating the three routes of convergence of bedforms identified and described in the text.

Examples of convergence of form along each of these three routes are described here.

Convergence by Route 1

(i) The two limbs of a *bifurcating furrow* of high relief (B_1) create between them, where they diverge, an *upstream-facing convex surface*, around which the limbs are wrapped in a manner similar to that of a *current crescent*. A *bifurcating furrow* may therefore contain within it a structure similar to an *obstacle mark* (B_2), and the two forms are convergent.

(ii) *Paired flutes* constitute a compound bedform (B_1) containing two intimately related component *flutes* separated by a central ridge. However, this arrangement is analogous to a simple *flute* (B_2) in which the ridge represents a type of *median ridge*. The only difference is that *median ridges* are rounded, whereas the ridge between *paired flutes* is relatively sharp.

(iii) *Open solution pits* (B_1, section 4.1a of the typology) are often similar in form to, and are therefore convergent with, small *flutes* (B_2). Furthermore, both bedforms are hypothesized to be potential precursors of *scallops*. *Open solution pits* are classified separately from *flutes* on the basis that they form part of a continuum of forms in limestone with other *solution pits*, and that they appear to arise independently of rock defects, unlike *flutes*.

(iv) Strongly curved *short furrows* and *parallel-sided furrows* (B_2, sections 1.2b and 1.2c of the typology) exist that are transverse at their upstream ends and become longitudinal at the downstream end, and that have a cuspate upstream rim in the transverse part. Such *furrows* are analogous to one-half of a *flute with median ridge* (B_1).

(v) In *pseudo-ripples with short furrows*, sinuous sharp crests (B_2) are often formed by, and consist of, the crests dividing contiguous *short furrows* that occur in the *troughs* (section 3.2a of the typology). Similar crests are contained within *compound parallel-sided furrows* (B_1), where they occur between the component *parallel-sided furrows* (section 1.2d of the typology).

Convergence by Route 2

(vi) *Breached potholes* (B_1) and *sharp-rimmed lateral potholes* (B_2, section 1.1e of the typology) are both incomplete *pothole* forms, often indistinguishable with any certainty, although they evolve along different paths. *Breached potholes* arise via the intermediate state of complete *potholes* (B_0), while *lateral potholes* were never complete.

(vii) *Obstacle marks* may be found in which the *obstacle* has become eroded down to the level of the surrounding bedrock outside the mark. These constitute a new bedform (B_1) similar to *flutes with median ridges* (B_2), but they have passed through the intermediate state of a typical *obstacle mark* (B_0).

(viii) Heavily *breached potholes* (B_1) retain only a small proportion of the circumference of the original complete *pothole* (B_0) and may overhang, thus resembling *overhanging concave surfaces* (B_2, section 1.5a of the typology), with which they are convergent.

Convergence by Route 3

(ix) *Longitudinal hummocky forms* (B_0) that attain high relief (B_{0M}) sometimes develop a sharp transverse crest that is typically not straight, but approximately parabolic in plan view, convex-upstream, enclosing a scoured hollow. This probably occurs because the relief of the *hummocky form* is sufficient to generate flow separation toward its downstream end. A *hummocky form* with such a crest (B_1) therefore contains within it a structure analogous to, and is convergent with, a *flute* (B_2), and may even develop a *median ridge* (Fig. 153).

(x) When *hummocky forms* and *SCHF* (B_0) attain high relief (B_{0M}), it is thought that they begin to behave as significant obstacles to the flow, creating conditions suitable for the development of *obstacle marks* (B_1) around topographic highs, and *transverse furrows* (B_1) within the *troughs*. Such *hummocky forms* are therefore convergent with a series of *obstacle marks* or *transverse furrows* that are developed directly (B_2).

(xi) *Directional scallops* (B_1, sections 2.1d and 4.2b) and *microripples* (B_2) have differing morphologies (crest steepness and planform), suggesting that they develop through different mechanisms. However, both being ripple-like forms, they are convergent to some extent. While the evolutionary path of *microripples* is uncertain, it is known that *directional scallops* arise via *flutes* or *flute*-like *solution pits* (B_0), which become conjugate (B_{0M}). Once they have become conjugate, the development of the *flutes* enters a new phase of modification and adjustment, eventually forming ripple-like *scallops*. The origin of *scallops* is discussed in a later section.

3. The Constructive Interference of the Flow Structures of Contiguous Bedforms

Although the flow structures within scoured forms remain at present largely a matter of conjecture, nevertheless from our current knowledge of them, it appears that when bedforms are contiguous, they are so arranged that their associated flow structures interact in a constructive and mutually reinforcing manner at their shared edge(s). This means that the time-averaged flow patterns in contiguous scoured forms meet in a concordant

Figure 153. Convergence by route 3, example (ix). *Longitudinal hummocky form* with a *flute*-like feature in its lee slope. The *flute* has a *median ridge*, on which the pen is resting. Borrow Beck, UK. Flow from top to bottom. Calcareous mudstone.

fashion at the common edge(s), and that the flow structures generated by or shed from one bedform impinge upon the second bedform in such a way as to enhance the flow structures that it requires for its development, and that it would generate anyway were it isolate. It may be argued that the existence of this principle is only to be expected, since if neighboring sculpted forms tended to interfere destructively, they either would display mutually avoiding behavior or would tend to destroy each other as the distance between them decreased. Therefore, the abundance of contiguous sculpted forms of a wide range of types effectively makes this principle self-evident. In passing, it may be added that the flow structures proposed by Maxson and Campbell (1935) as being present in conjugate *flutes* imply destructive interference between adjacent individuals and are therefore highly unrealistic. They were in fact subsequently found to be incorrect by the experiments of Allen (1971a).

The principle of constructive interference applies to (1) contiguous individuals of the same or of different types, (2) neighbors in conjugate assemblages and individuals within space-filling trains of bedforms (e.g., *scallops*), and (3) neighboring components within a compound individual, provided that in all these cases the shared edge(s) are sharp crests. The presence of sharp crests shows that the individual bedforms or component features are active centers of erosion, and that they and their associated flow structures are indeed contiguous with and are interacting with neighboring forms. The interpretation of sharp crests and of flow structures within bedrock bedforms is discussed further in a later section. These three different applications of the principle of constructive interference are discussed in more detail here.

(i) The principle of constructive interference requires that where the vortices associated with neighboring features impinge upon each other, they should be counterrotating. This is satisfied, for example, in the case of *flutes* lying side by side, or with contiguous *parallel-sided furrows* generating a *bladed form*. The presence of a sharp crest between two contiguous sculpted forms implies flow separation at the crest, which in turn implies that the flow structures on either side of the crest are counterrotating. It can be seen therefore that the principle of constructive interference follows largely from the existence of contiguous bedforms divided by sharp crests.

(ii) As an example of constructive interference in a space-filling train of bedforms, one may take *en echelon three-dimensional scallops* (Fig. 129). The abundance of the en echelon arrangement suggests that it is in some way a preferred or stable state. It is easy to see why this should be the case if bedforms tend toward a state of constructive interference. Given that a *three-dimensional scallop* assemblage is a collection of peaks and parabolic scour hollows, then a very constructive arrangement for these features is an antiphase (en echelon) one in which the peaks closely approach or impinge upon the upstream ends of the next depression downstream and in which the peaks therefore act as "defects" for the scour hollows, shedding vortices into them and enhancing their development. In this manner, the depressions are analogous to isolate *flutes* not only in shape, but in function. This theory also explains why the in-phase arrangement of peaks is rare; it leads to destructive interference in which the upstream peak tends to destroy the downstream peak. It also suggests a mechanism for the origin of secondary *scallops* within a primary network, that of their nucleation in the wake of existing peaks, and a mechanism by which *scallop* assemblages may tend toward an en echelon arrangement.

(iii) All compound bedforms in which the component *concave features* are divided by sharp crests exhibit constructive interference between the components. In *hierarchical potholes* (Fig. 18) and *hierarchical lateral potholes* (Fig. 30), for example, observation of the internal morphology indicates that systems of related counterrotating vortices are set up within the component *potholes* when active. Flow in these vortices meets concordantly at the high-angle crests (see later section for a discussion of low-angle and high-angle crests) dividing the secondary *potholes* from the primary one, and the vortices therefore interact in a constructive manner. In *nested obstacle marks*, the attached horseshoe vortex associated with one *obstacle* is enhanced by being in the lee of the next *obstacle* upstream. Further examples include paired bedforms and some types of *furrow complexes*, such as the *yin-yang* type (section 1.4b of the typology), all of which display distinctive geometrical arrangements implying very close associations and constructive interactions between the component features. The flow structures within compound bedforms, whose component *concave features* are not separated by sharp crests, and the applicability of the principle of constructive interference to them, remain much more a matter of conjecture.

4. The Prevalence of Sharp Transverse Crests

SCHF of one type or another are very common in bedrock channels across a range of lithologies. The development of *SCHF* is favored by a homogeneous substrate or a *concordant plane bed.* Although extensive developments of *pseudo-ripples* and *pseudo-dunes*, which appear to require large areas of relatively smooth *plane bed* and uniform flow, are rare, a *restricted development of SCHF* is almost ubiquitous. This suggests that a surface with repeated topographic singularities (which sharp transverse crests represent) is a natural state to which bedrock channels tend, and such a surface will become extensive and quasi-periodic if conditions allow.

Because *SCHF* occur in a range of situations and apparently arise through a variety of mechanisms (which are discussed below) with broadly similar results, this is really just a further example of the convergence of form. However, *SCHF* are so common in bedrock channels that they are highlighted by being discussed separately. Of all the examples of convergence of form, they are by far the most important. Longitudinal sharp crests also occur (e.g., *bladed forms* between contiguous *longitudinal furrows*), but they are much rarer and are effectively a by-product of the development of conjugate *longitudinal furrows*. *SCHF* do not require the presence of any other bedforms and are apparently an independent form capable of arising spontaneously in bedrock surfaces without obvious defects.

The Origin of Two Types of Sharp-Crested Transverse Features

Sharp-Crested Hummocky Forms and Scallops: A Comparison

These two types of sharp-crested transverse features have been separated throughout this report. This is because, despite some obvious similarities and the continuum that apparently exists between them in both scale and form, the two are in general distinct forms in the field sites studied. The main distinctions between the two bedforms are described here:

(i) The wavelength of *SCHF* is typically an order of magnitude larger than that of *scallops*, although there is a complete gradation between *scallops* and *SCHF* in scale.

(ii) Although the planform pattern of *scallops* is variable, highly two-dimensional forms are rare, and assemblages tending toward the three-dimensional end of the spectrum are the norm. This three-dimensional nature arises through (1) the crests consisting of angular, downstream-pointing sections and parabolic, convex-upstream sections that enclose a scour hollow, between the angular ones, and (2) the generally short lengths of crest which exist between crest nodes, with crests exhibiting a high degree of connectivity in the longitudinal direction. The crests of *scallops* tend not to fade out laterally, but to terminate at intersections with other crests (i.e., nodes). *SCHF* on the other hand, tend to be more two-dimensional with crests of low sinuosity that lack angular sections. They generally have irregular perturbations in plan view and no clear pattern such as that described for *scallop* crests. There is very low longitudinal connectivity between crests and few crests nodes. Crests tend to fade out laterally without coming into contact with the neighboring upstream and downstream crests.

(iii) *Scallops* are not associated with *longitudinal furrows*, but *SCHF* often are (section 3.2a of the typology). Where a *pseudo-ripple* crest meets a *longitudinal furrow* (e.g., at the end of a *short furrow* or as a crest within a *regular compound furrow*), this tends to cause a downstream-pointing perturbation in the crest. Where *longitudinal furrows* are present, therefore, the downstream-pointing crest sections in *SCHF* tend to occur at depressions and are curved (Fig. 105A), rather than occurring as peaks and being angular as they are in *scallops* (Fig. 128). Conversely, *SCHF* have not been observed to form *longitudinal ridges* through constructive interference of the crests, as *scallops* have (section 4.3d of the typology).

(iv) Whereas *scallops* always completely fill the space available within an assemblage, within a field of *pseudo-dunes* there may be large spaces with no crests in them. In fact, *SCHF* range from isolated crests to fields of tightly packed crests illustrating space filling.

(v) Whereas *SCHF* appear to arise through several mechanisms (see below), *scallops* are generally regarded as forming through the conjugation of *flutes* and *flute*-like *solution pits*.

Nevertheless, it seems likely that *pseudo-dunes* and *pseudo-ripples* are closely related to *scallops*, in view of the continuum that exists between them. A continuum in morphology is demonstrated by areas in which *SCHF* take on a planform very similar to *three-dimensional scallops*, but without any reduction in wavelength. This occurs, for instance, in Nam Mae Khlang where, although the *pseudo-dunes* show only a patchy development, in places it is highly three-dimensional with parabolic scour hollows and angular downstream-pointing crest sections (Fig. 88). The Rivers Dee and Wharfe illustrate continua in both size and morphology, where examples of *pseudo-dunes* and *pseudo-ripples* occur with wavelengths varying from several decimeters down to only a few centimeters. These changes are coupled with an increasingly three-dimensional crest planform.

Furthermore, the longitudinal profiles of *pseudo-dunes* and *scallops* are similar, being convex-up on the stoss sides and concave-up on the lee sides, with a relatively sharp crest in between. It is known that *scallops* are a downstream-migrating bedform in which the topographic and erosion rate profiles are out of phase (the vertical erosion rate increasing with slope), and in which flow separation occurs at the crests (Allen, 1971a; Blumberg and Curl, 1974; Allen, 1982: p. 219–222, 242–245). Given the similarity of the profiles and given also that mass transfer by dissolution in soluble materials and abrasion in cohesive insoluble materials are analogous (Allen, 1971a; see next section), the characteristics of *scallops* described above may reasonably be assumed to be true also of *SCHF*.

The Origin of Scallops

It has been widely proposed, on the basis of theoretical considerations, that the origin of *scallop* patterns in limestone lies in the ability of a bed to respond by mass transfer to the flow effects induced by random defects in the bed (Coleman, 1949; Allen, 1971a, 1982: p. 269–270; Blumberg and Curl, 1974; Selby, 1977; Ford and Williams, 1991: p. 305; Slabe, 1996: p. 31). This is the so-called "defect theory" of *scallop* origin. Its rival, the "passive bed theory," assumes that *scallops* are effectively the imprints of flow structures that exist independently within the boundary layer (Allen, 1971a). The growth of *scallop* patterns in plaster of Paris through the conjugation of *flute* marks arising at random defects was observed directly by Allen (1971a, 1971b) and Goodchild and Ford (1971), thus supporting the defect theory. These *flute* marks were similar to those formed in mud by localized scour in the lee of bed defects (Allen, 1971b), and confirmed the postulate of Allen (1971a) that mass transfer by dissolution (in plaster of Paris) is analogous to that induced by abrasion in a cohesive substrate.

The origin of *scallops* through the conjugation of *flutes* is also supported by field morphological evidence, as described here:

(i) Molds (i.e., negative surfaces) taken of *scallop* assemblages (Fig. 164) show a feature of the bedform that is not readily appreciated in its original positive form, the appearance of which is dominated by the ripple-like crests. This is the fact that *scallop* assemblages contain a preponderance of parabolic, *flute*-like scour hollows. Indeed, such molds are similar to the molds of conjugate *flutes* formed in mud and preserved on the soles of sandstone beds in turbidite deposits. This was found by Allen (1971a) with molds of his experimental *scallops* in plaster of Paris and has been confirmed by a mold taken by the authors of a natural *scallop* assemblage in limestone at Sleightholme Beck. This observation of course implies that a range of scour forms, from isolate and conjugate *flutes* to three- and two-dimensional ripple-like forms, should also be found in mud deposits, just as they are in limestone. This is indeed the case; Dzulynski and Sanders (1962: Plates IVB and VA therein) describe from turbidite deposits both two-dimensional ripple-like forms and bedforms transitional between these and conjugate *flute* assemblages.

(ii) The pattern of *three-dimensional scallops*, dominated by convex-upstream parabolic crest sections enclosing hollows and by angular downstream-pointing crest sections between them, is what would be expected from the complete conjugation of an assemblage of *flutes* (Fig. 7), and indeed some examples of conjugate *flutes* do have an appearance that is transitional with *three-dimensional scallops*.

(iii) The en echelon arrangement observed in some *flute* assemblages (e.g., Fig. 39 of this report; Maxson, 1940: Fig. 1 therein) mimics that observed in *three-dimensional scallop* assemblages and suggests a further mechanism, in addition to that relating to the constructive interference of the flow patterns, for the origin of en echelon arrangements of *three-dimensional scallops*, i.e., that such arrangements are inherited from antecedent flute assemblages.

(iv) New *scallops* can be nucleated within an existing assemblage at "defects" within the rock, i.e., they are nucleated in the same manner as *flutes*. This observation is described in more detail below.

However, the defect theory of *scallop* origin suggests that *scallop* spacing, and therefore wavelength, is dependent upon the spacing of the defects at which the antecedent *flutes* originated. This conclusion contradicts the field observations of Curl (1966) and the experimental results of Goodchild and Ford (1971) and Blumberg and Curl (1974), who found that *scallop* wavelength is inversely proportional to flow velocity. A similar relationship was observed by Ashton and Kennedy (1972) for the wavelength of ripples on the underside of ice covers. Fortunately, the experiments of Allen (1971a) provide quantitative evidence that suggests a resolution of this issue. Although he did not run extremely long experiments, the results showed that despite an initial control of *scallop* wavelength by the spacing of defects at which *flutes* originate, as the age of the *scallop* assemblages increases, the wavelength appears to become controlled by flow velocity. Using this result, Allen (1971a) proposed a "modified defect theory," in which secondary marks are generated by primary ones within mature assemblages of isolate *flutes*, and within fully conjugate assemblages (i.e., *scallops*). This has the effect of modifying the wavelength of *scallops* from that which would result from the spacing of the original defects.

In accordance with the evidence discussed above, the formation of *scallops* is envisaged as a two-step process:

(i) A field of initially isolate *flutes* enlarge, interfere, and become fully conjugate, undergoing a transformation in planform, and creating a *scallop* assemblage. Thus, the initial wavelength of the *scallops* depends largely on the spacing of the original defects.

(ii) The new *scallops* are modified by mutual interaction, during which the assemblage adjusts to the prevailing flow conditions in a feedback process. This process of modification results in changes in both the planform pattern of the crests and the statistical properties of the *scallop* assemblage (such as mean wavelength). The modification occurs by the suppression of certain crests and associated decay of mature *scallops*, and by the continuous generation of new, secondary *scallops*. Both of these processes have been observed experimentally (Allen, 1971a; Blumberg and Curl, 1974). The existence of this process of modification and adjustment to the flow is also suggested by the gradation between two- and three-dimensional types, observed in the field at a single site, indicating the evolution of the planform patterns of *scallop* assemblages, and by the observed dependence of *scallop* wavelength on flow velocity. Eventually, the *scallop* assemblage may achieve a state of equilibrium with the flow. This equilibrium may be a static one, resulting in stable bed profiles, such as those hypothesized by Curl (1966) and produced artificially by Blumberg and Curl (1974), or a dynamic one in which individual *scallops* are continuously created and destroyed, but in which the statistical properties of the assemblage are stable, as was also conceived by Curl (1966).

Scallop wavelength may also be influenced by lithology. The physical structure of the rock, in terms of its heterogeneity, affects the roughness of the eroded surface, which in turn reduces the average wavelength of the *scallop* assemblage by nucleating new *scallops* at "defects" (Goodchild and Ford, 1971). This conclusion is supported by detailed observations of individual *scallops* at Sleightholme Beck. At this site, many "defects" exist in the rock in the form of fossil fragments up to

~10 mm across. Some of these defects stand proud of the surrounding rock and generate localized scour, which begins as a small *current crescent* around the fossil, but which grows into a small *flute* and ultimately becomes a *three-dimensional scallop*, blending in with the surrounding assemblage. The outward growth of the *scallop* in all directions results in the original defect becoming closer to the center of the depression. Fossils standing proud near the center of *scallops* are reduced to narrow, delicate spire-like shapes, which are highly eroded relics of the original fossil. There are also many *scallops* in which a fossil is found at the center, but in which the fossil is worn almost completely smooth. It seems that at this site, fossils do initiate localized erosion around them, but that at a certain stage in its growth, the depression becomes self-sustaining and develops independently of the fossil, becoming assimilated within the surrounding *scallop* pattern. The fossil, now near the center of the depression and no longer driving the hydrodynamics within it, finds itself in a local erosion maximum and is quickly eroded away, leaving the depression smooth. However, Slabe (1996: p. 14, 17) and Ford and Williams (1991: p. 305) suggest that if the bed surface is too rough or heterogeneous, the resulting turbulence field may be too chaotic and may prevent the formation of *scallops*. This conclusion is also supported by observations from Sleightholme Beck, where the substrate is highly heterogeneous in patches, and in those areas, the rock surface is rough and devoid of *scallops*.

The role of defects and *flutes* in the development of *scallops* seems quite clear. However, an alternative mechanism also exists. In the Rivers Dee and Wharfe, which flow over limestone, there are only incipient and poorly developed *scallops*. This may be because the rock appears to lack the large and abundant fossil fragments that might act as defects, or it may be because the gradient of these rivers is much less than that of Sleightholme Beck, and therefore the flow velocity is insufficient to generate the intense scour necessary to produce *flutes* at defects. Nevertheless, in the Rivers Dee and Wharfe there are numerous examples of assemblages of *space-filling fluvial solution pits* of the *elongated* and *open* varieties. Space-filling assemblages of isolate *solution pits*, especially those with en echelon arrangements, can be reasonably explained in only two ways:

(i) By primary *solution pits* generating external closely spaced but separate secondary marks. Allen (1971a) described several ways in which secondary marks may be generated from primary ones, but these all operated either internally within the primary mark (and thus acted to subdivide it) or at the crest between two conjugate primary marks. Kor et al. (1991) and Shaw (1996) describe space-filling en echelon arrangements of sichelwannen from Georgian Bay, Ontario, and also invoke an ability of primary sichelwannen to generate secondary forms to explain the patterns, but according to these authors the bedforms were interfering, and therefore partly conjugate. A type of sedimentary bedform, the rhomboid ripple mark (Hoyt and Henry, 1963; Otvos, 1965), also develops an en echelon arrangement, but in this case the pattern is more regular. There are two mechanisms that could be responsible for the generation of external secondary *solution pits* in a characteristic position relative to the primary one. These are (1) the nucleation of new *pits* at the edge of the turbulent wake downstream of an existing one, and (2) the nucleation of new *pits* under the oblique shock wave emanating from an existing *pit* in shallow, supercritical sheet flow down the lee slope of a *hummocky form*, where *space-filling solution pits* are very common. Such shock waves have been observed by the authors at low stage. In littoral sedimentary environments, oblique shock waves produce erosional forms known as rhomboid rill marks (Otvos, 1965). The less common longitudinal alignments of *solution pits* apparently arise through nucleation of secondary pits within, rather than at the edge of, the turbulent wake of an existing one.

(ii) By the existence of a strong driving force for the nucleation and growth of new *solution pits*, which acts everywhere on a particular surface, but which is suppressed in the immediate vicinity of existing *solution pits*. Pitting will therefore be intense, and *solution pits* will fill the space available but will show a tendency for mutual avoidance. This would produce a quasi-close-packed arrangement of *pits* in which en echelon patterns are common.

In some places, *space-filling solution pits*, though normally isolate, do become conjugate, and in so doing resemble incipient *three-dimensional scallops*. Therefore, an alternative mechanism to *flute* formation at random defects for the generation of *directional scallops* is that of the complete conjugation of *space-filling solution pits*. The distinction is nontrivial because *flutes* and *solution pits* are different both morphologically and genetically in that the development of *solution pits* appears generally not to be dependent upon the existence of defects. The *space-filling solution pits* are not distributed randomly and arise either through the generation of secondary pits from primary ones or through a mechanism where new pits are not nucleated close to existing ones. The tendency for en echelon development in *space-filling solution pits*, like that seen in some *flute* assemblages, also mirrors that in *three-dimensional scallop* assemblages. In addition, at one site where *longitudinal ridges* of the variety formed through in-phase, longitudinal alignments of *scallops* (section 4.3d of the typology) are common, the *scallops* grade laterally into areas of isolate *solution pits* where longitudinal rows of pits are also common.

The above discussion has considered only *directional scallops*. In the generation of the nondirectional variety, which arises in conditions lacking a strong unidirectional flow, *flutes* cannot be the agent for the inception of *scallops*, and *shallow circular solution pits* are an obvious and strong candidate. Indeed, at several sites in the River Dee, *nondirectional scallops* were observed to grade laterally into isolate *shallow circular solution pits* in places such as slack water areas and flow separation cells (Fig. 154).

The Evolution of Directional Scallop Planforms

The idea of the evolution of *directional scallop* planforms was mentioned briefly in the two-step process of formation described above. The existence of the process of planform evolution is actually necessitated by the fact that the complete conjugation of a field of *flutes* or *solution pits* will produce an assemblage of highly *three-dimensional scallops*, and yet observed *scallop* assemblages illustrate a range of planform patterns from three- to two-dimensional. Curl (1966) and Blumberg and Curl (1974) view the formation of three-dimensional scallops as the breakdown of an ideal two-dimensional form, due to unsteady flow conditions. The geomorphological evidence cited above, however, suggests an alternative line of reasoning, namely that *two-dimensional scallops* can be thought of as the fullest development of a process that leads from isolated scours to *scallops* of a highly three-dimensional kind via conjugation, which *scallops* then undergo a process of evolution by which the crests become straighter and more directly transverse. The number of crest intersections decreases, reducing the longitudinal connectivity of crests, and individual depressions become of larger lateral extent, i.e., they become *intermediate scallops*. This process can be seen to some extent in the plaster of Paris experiments of Allen (1971a: Figs. 70 and 71 therein), in which the most mature *scallop* assemblages contain some *intermediate scallops*, whereas younger assemblages do not. Under special conditions that are at present unclear, this evolution process leads ultimately to *two-dimensional scallops*. This general scheme of development is supported by the inability of experimental studies in plaster to generate *two-dimensional scallops*, in spite of imposed steady flow conditions. Blumberg and Curl (1974), for example, found it necessary to induce them by carving transverse grooves into the model.

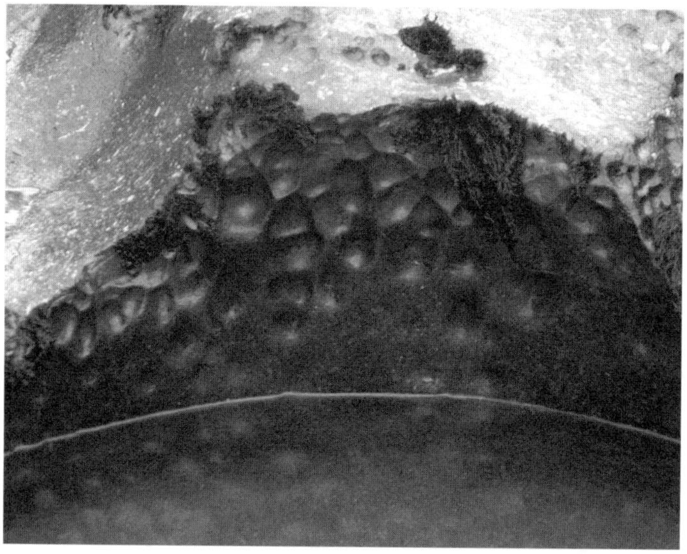

Figure 154. *Nondirectional scallops* forming through the conjugation of *shallow circular solution pits* in the lee face of a *pothole*. In the lower part of the *pothole* lee face, some *solution pits* remain only partially conjugate or isolate, but in the upper part they are fully conjugate and can be classed as *nondirectional scallops*. River Dee, UK, looking upstream. Limestone. The field of view is ~30 cm.

The Origin of Sharp-Crested Hummocky Forms

Unlike the generation of *scallops*, an analysis of the various morphologies of *SCHF* (*sharp-crested hummocky forms*) and their associations with other bedforms suggests that they may be formed in several ways, depending upon their environment. These mechanisms, which are illustrated schematically in Figure 155, appear not to be mutually exclusive and probably operate alongside each other. Some mechanisms act to produce *pseudo-ripples* or *pseudo-dunes*, while others generate isolated crests. They include the following:

(i) *Formation under standing waves and hydraulic jumps* in the case of *runnels with SCHF* (section 3.2b of the typology; Fig. 155A). Standing (unbroken) waves and hydraulic jumps are maintained within *runnels with SCHF* over a range of low to moderate flows and are drowned out only at high flows. The standing waves and/or hydraulic jumps are associated with the bed topography of the *runnel.* They are often in the form of oblique or V-shaped shock waves, and it is worth noting that the *SCHF* in some *runnels* are also of this form. That V-shaped and oblique shock waves are virtually ubiquitous in *runnels*, even those without *SCHF*, and in bedrock channels generally, suggests that these flow features are often the cause of *SCHF* within *runnels*, and that the bed is responding passively to them by developing a wavy profile. Once *SCHF* have begun to develop, standing waves and/or hydraulic jumps will be fixed over them, enhancing their growth and producing a positive feedback effect. Although the precise way in which the sharp crests are generated is not clear, it must involve the nonuniformity of the flow and its repeated acceleration and deceleration and associated variations in turbulence intensity. Turbulence intensity is known to increase within a diverging (but unseparated) flow (Ashton and Kennedy, 1972), for instance, and hydraulic jumps are visibly extremely turbulent. These properties of the flow have the potential to produce repeated variations in erosion rates. Where turbulence intensity is higher, the movement of water is extremely vortical, streamlines have a small radius of curvature, and suspended grains are more likely to become decoupled from the flow and flung at the channel boundary (the coupling of streamlines and suspended particles is discussed further below). Tinkler (1993) also noted the role of hydraulic jumps in localizing abrasion, and Tinkler and Wohl (1998a) described a *runnel* from Cooksville Creek, Ontario, which possessed an undulating bed, and which is occupied by standing waves of a wavelength similar to the bed undulations.

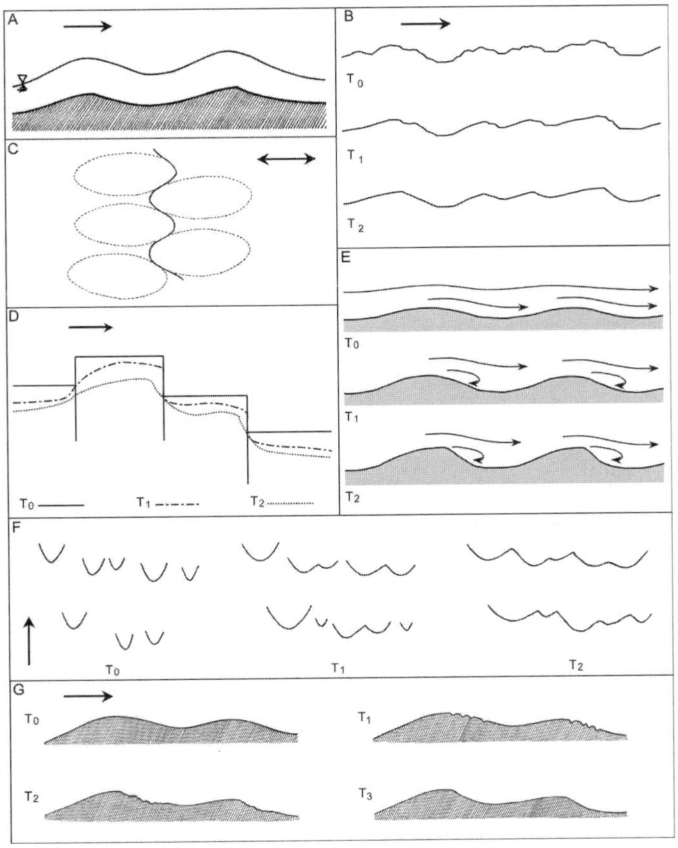

Figure 155. Proposed modes of formation of *sharp-crested hummocky forms*. See text for explanation. Arrows indicate current direction. T_0, T_1, etc., refer to time sequence.

Figure 156. This photograph showing a *partially abraded surface* was taken close to that in Figure 90. Abrasion here is present over most of the bedrock surface and has resulted in the formation of *sharp-crested hummocky forms*. Gayle Beck, UK. Flow from top left to bottom right. Limestone. The scale is 60 cm long.

(ii) *The abrasion and smoothing of irregular surfaces* in which the bed irregularities generate flow separation, owing to their relief and small radius of curvature (Fig. 155B). Such surfaces become first *partially abraded surfaces* with a sharp boundary between the smooth (highly abraded) and rough (lightly abraded) areas, the boundaries representing lines of flow separation, and eventually become *SCHF* once abrasion has smoothed the whole surface. The lines of flow separation remain as sharp crests (Fig. 156).

(iii) *The development of crests between contiguous longitudinal furrows* (Fig. 155C). In the case of *SCHF* in which *short furrows* are present in the *troughs*, the transverse crests, or sections of them, are sinuous and are often formed by, and are identical to, the crests dividing contiguous *furrows*. (Fig. 105).

(iv) *The abrasion of rock outcrops that are bounded by planar structural surfaces* of different orientations (i.e., joint, bedding, and cleavage surfaces, etc.) (Fig. 155D). Edges formed at the intersection of these surfaces generate flow separation. After an extensive amount of abrasion, the outcrop as a whole will become rounded and smoothed, but such edges may be preserved as sharp crests on the outcrop, continuing to generate flow separation, even though they may have migrated significantly from the position of the original edge, and become sinuous (Fig. 157).

(v) *The generation of flow separation* on the lee sides of smooth *hummocky forms* (or any other positive feature) that attain sufficient relief and a sufficiently small radius of curvature (Fig. 155E). Flow separation will then create a discontinuity in the erosion rate profile, which will in turn generate a sharp crest at the point of separation. This process was observed in the formation of sharp crests in ripples on the underside of ice covers (Ashton and Kennedy, 1972). Hancock et al. (1998) also proposed that low-amplitude, rounded *hummocky forms* are precursors to higher-relief *SCHF.*

(vi) *The conjugation of fields of large flutes*, i.e., the same manner as that in which *scallops* are formed. This mode of formation is suggested by the occasional similarity of planform that occurs between *pseudo-ripples* and *three-dimensional scallops* (Fig. 88). The difference between the two types of bedforms in this case is essentially one of scale and the fact that the three-dimensional *pseudo-ripples* are associated with other types of *pseudo-ripples* nearby. *Restricted developments of SCHF* can be formed in a similar manner from only a few *flutes*, or in some cases *incipient potholes*, which happen to lie close to each other approximately in a line transverse to the flow. The growth and interference of such features leads to their rims becoming connected laterally, forming a continuous sharp, transverse crest (Figs. 155F and 158). Crests resulting from this mechanism will contain within them,

Figure 157. *SCHF* arising through the abrasion of outcrops bounded by structural planar surfaces, and trending parallel to the strike of the substrate. These sculpted forms occur in a *discordant plane bed* with steeply dipping bedding planes that strike across the channel. Resistant strata form bedrock ribs, such as the one on which the figure (Hiroshi Ikeda) is standing. The front edges of these ribs are subject to abrasion and modification of the initial structural surface. The front edges migrate downstream but remain as sharp edges. Sichiri-gawa, Japan. Flow from left to right. Interbedded mudstone and fine-grained sandstone.

in plan view, sections of parabolic form that are convex-upstream, and angular downstream-pointing regions. This process is facilitated by the fact that *flutes* can be initiated at "defects" that consist of mere changes in orientation of the surface.

(vii) *The solution pitting of the lee sides of rounded hummocky forms in limestone* (Fig. 155G). When solution pitting is intense, in terms of either the size of the *pits* or the number of them, coalescence of the *pits* results in the development of *convex surfaces with steep lee faces* (section 3.3a of the typology). These are not regarded as types of *SCHF*. Less intense pitting, however, which is common on the lee faces of *hummocky forms*, can result in the development of sharp, transverse crests on them with little additional alteration. This appears to occur through the coalescence, in the first instance, of the leading (i.e., upstream) row of *solution pits* (Fig. 159). Crests resulting from this mechanism will also initially contain within them small-scale curved sections and angular downstream-pointing regions, but these may be straightened out in time.

The Development of Sharp Edges in General

The Phenomenon of Sharp Edges in Bedrock Bedforms

The discussion of sharp-crested features has so far considered only trains of transverse ripple- and dune-like forms. However, these are just part of the wider phenomenon of the development of sharp edges in sculpted forms in general. Sharp edges include the crests dividing contiguous bedforms, the rims of some isolate bedforms, and internal sharp crests in many compound forms. They vary in their orientation, steepness, and relationship with

Figure 158. *SCHF* arising through the coalescence of parabolic *flutes* in a lateral direction and the straightening of the rims. In the center foreground, isolate *flutes* can be seen; elsewhere, they appear to have coalesced into *SCHF*. Nahal Hatira, Israel. Flow from bottom right to top left. Limestone. Camera bag, 20 cm wide, for scale.

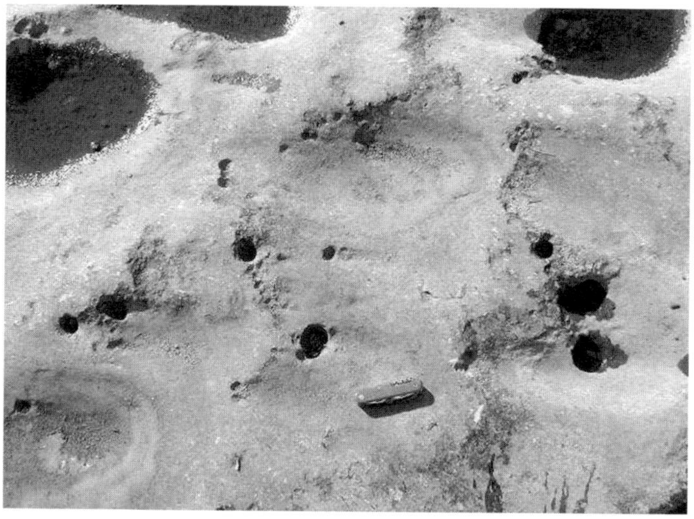

Figure 159. *SCHF* arising through solution pitting on the lee sides of initially smooth *hummocky forms*. Gayle Beck, UK. Flow from left to right. Limestone. Penknife for scale.

other sculpted forms. The subject of the recognition of *breached* and *lateral potholes* has already been discussed (sections 1.1d and 1.1e of the typology). Zen and Prestegaard (1994) cite the rounded rims of *lateral potholes* as evidence that they are not *breached potholes*, which implies that sharp rims are evidence of breaching. Observation of a range of bedforms, including many *potholes*, however, suggests that this is an incorrect assumption, and that sharp edges commonly form the boundaries of scoured features that have almost certainly never been breached, and that the development of sharp edges is a widespread phenomenon.

A Criterion for the Development of Sharp Edges

At a fundamental level, two deductions may be made about sharp edges in sculpted forms:

(i) Abrasion, while active on the surfaces to either side of the sharp edge, is absent from the edge itself, otherwise it would be rounded. The flow must therefore separate from the boundary at the edge, in order to obviate abrasion there. The sharp edge represents the intersection of two surfaces of different orientations that are actively eroding, though not necessarily at the same rate, and that are retreating toward each other as a result of their difference in orientation.

(ii) The sharp edge, being a singularity in the topographic profile and therefore also in the erosion rate profile, must at least represent an abrupt change in the flow pattern immediately above the surface, in other words a boundary between two temporally and spatially persistent coherent flow structures.

The presence of strongly convergent skin friction lines on either side of a sharp edge is a requirement of the first deduction, regarding the separation of the boundary layer, and satisfies the second deduction. Strongly convergent skin friction lines would therefore seem to be a necessary and sufficient criterion for the development of sharp edges. This was in fact found to be the case in the detailed skin friction maps of *flutes* and *scallops* obtained experimentally by Allen (1971a) and is discussed further below. It is also apparent that the surfaces on either side of a sharp edge are experiencing different erosion regimes in response to distinct but interacting and related flow structures. The principle of constructive interference of the flow patterns of contiguous bedforms discussed earlier suggests that this interaction always occurs constructively so as to reinforce the flow structures.

In contrast, the absence of a sharp edge in an actively abrading surface (e.g., *hummocky forms* and many *lateral potholes*) implies the absence of discontinuities in the distribution of erosion rates and therefore the absence of abrupt changes in flow pattern over the surface. It therefore implies either that flow separation is absent (which is likely in the case of *hummocky forms*) or that the surface is in a highly turbulent environment subject to the frequent passage of vortices and associated ejections and sweeps, such that a boundary layer is not established and flow separation cannot occur at a fixed position (which is likely in the case of *lateral potholes* with rounded rims).

Types of Sharp Edges

Sharp edges in bedforms arise in a variety of situations and take various forms. It is proposed that sharp edges may be subdivided into active and passive types:

(i) Active sharp edges are those at which flow separation occurs and at which skin friction lines are strongly convergent, as discussed above. These edges interact with the flow (by generating flow separation), and the flow then acts to cause erosion that enhances the sharp edge in a positive feedback cycle. Sharp edges in such systems may be divided into two types, based on the angle by which the orientation of the surface changes across the edge (Fig. 160). High-angle sharp edges are those in which the surface orientation changes by more than 90° across the edge, and where fluid approaching the edge within the flow structures on either side meets relatively obliquely, and therefore in a concordant fashion, at the edge when viewed in the plane normal to that edge (Figs. 160B and 160C). In contrast, low-angle sharp edges are those in which the surface orientation changes by less than 90° across the edge, and where fluid approaching the edge from either side is

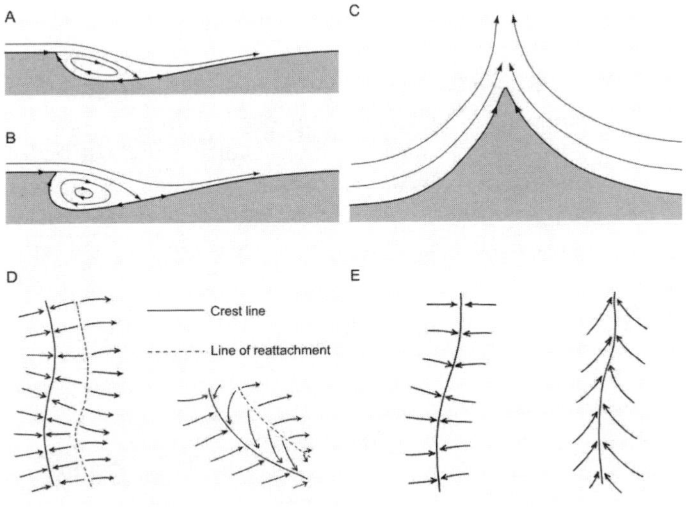

Figure 160. Types of sharp edges and their inferred associated flow patterns. A and B: Cross sections through low- and high-angle sharp edges respectively, forming the rim of isolate *concave features* (A modified after Allen, 1971a). Free streamflow from left to right. C: Cross section through a high-angle edge forming the crest between two contiguous *concave features*. Free streamflow direction is arbitrary. D: Potential skin friction lines for the edges shown in A and B for the directly convergent case of a purely transverse crest (left) and the obliquely convergent case of crests of other orientations (right). Free streamflow from left to right. E: Potential skin friction lines for the edge shown in C for the directly convergent case (left) and obliquely convergent case (right). Free streamflow direction is arbitrary. Directly convergent skin friction lines indicate a roller-type vortex; obliquely convergent skin friction lines indicate helical vortices.

traveling in nearly opposing directions when viewed in the plane normal to the edge, and therefore meets at the edge in a discordant fashion when viewed in this plane (Fig. 160A). Most sharp edges are probably active, and they exist in one of three different situations, illustrated schematically in Figure 161A:

- At the rim of isolate *concave features*. Many scoured hollows eroding rapidly into a surface in which much slower general erosion is occurring have cuspate rims. An established or partially established boundary layer separates from the boundary at the rim and fluid moves in generally helical paths in attached vortices within the bedform (e.g., Allen, 1971a). Examples include *flutes* (Fig. 161A [i]), *current crescents*, *overhanging concave surfaces*, and *sharp-rimmed lateral potholes*. This type of sharp edge may be low-angle or high-angle. For instance, during the development of an *overhanging flute* (section 1.2a of the typology) from one which does not overhang, the rim changes from a low-angle to a high-angle edge as it begins to overhang at the proximal end. A similar phenomenon seems to occur with overhanging *current crescents*.
- At the crest between contiguous *concave features*. Any two or more actively eroding *concave features*, if they are contiguous and interacting, and if they remain as distinct entities rather than coalescing and uniting, may develop sharp crests dividing them. The *concave features* may be contiguous individuals or neighboring components within a compound individual. *Bladed forms*, for instance, are high-relief longitudinal ridges with sharp crests formed between contiguous *longitudinal furrows*. In this respect, *bladed forms* are analogous to the keels of wedge-shaped snow and sand drifts described by Whitney (1983), which were sharp edges generated by "simultaneous off-sweep from two sides." Examples from compound individuals include the sharp crests within *regular compound parallel-sided furrows* (Fig. 161A [ii]) and those between the primary and secondary *potholes* in *hierarchical potholes*. Crests between contiguous *concave features* may be low- or high-angle.
- At the crests of positive features of sufficiently high relief and small radius of curvature, such that the boundary layer separates at the crest. The crests are generally low-angle. The positive features may be structurally or hydrodynamically controlled, and they may be isolated or part of a quasi-periodic train of bedforms. Examples include some *partially abraded surfaces*, *SCHF* (Fig. 161A [iii]), and *scallops*.

(ii) Passive sharp edges are those that arise incidentally through the action of a process that does not necessarily lead to the development of a sharp edge, and in which the sharp edges, once developed, play no active part. The discussion above relating to convergent skin friction lines and flow separation does not apply to passive sharp edges. They are probably not common but may occur in the following ways (see Fig. 161B):

- Through surface planation in an area of inactive bedforms. Where general surface lowering (planation) is occurring on a surface into which deep bedforms, such as *potholes*, have been eroded, sharp rims may be created around the apertures of the bedforms. This appears to be happening to *coalesced potholes* on the banks of Nam Mae Chaem (Figs. 161B [i] and 162) and is the result of the removal of the originally rounded rims (of the *potholes*) by erosion due to a process other than

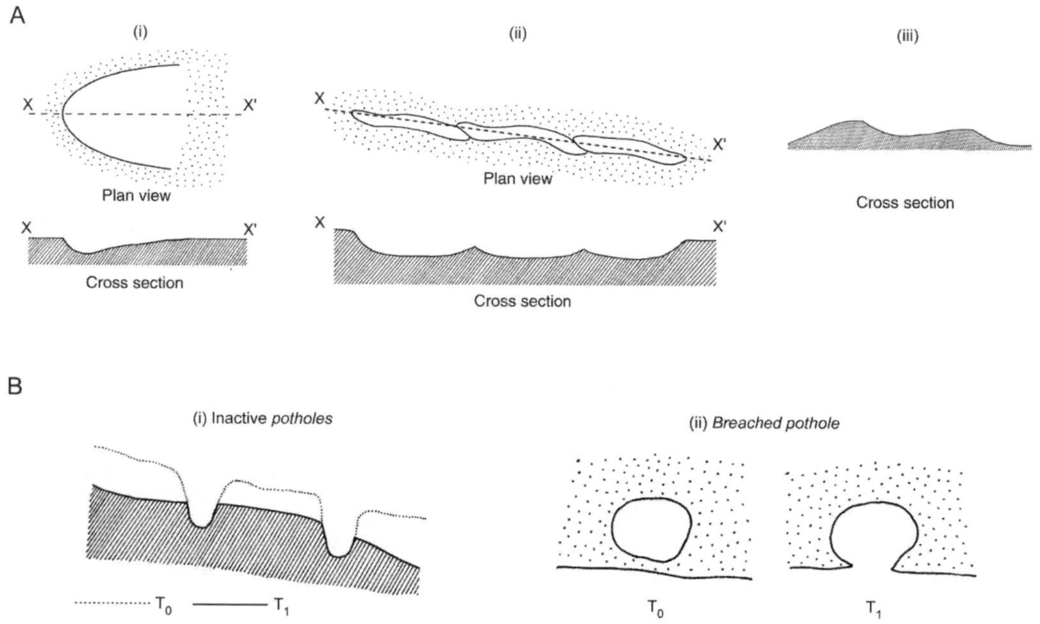

Figure 161. The origins of active and passive sharp edges. A: Active sharp edges form (i) at the rim of isolate *concave features*, (ii) at the crests between contiguous *concave features*, and (iii) at the crests of positive features of sufficient relief. B: Passive sharp edges form (i) through surface planation, and (ii) through breaching. See text for explanation.

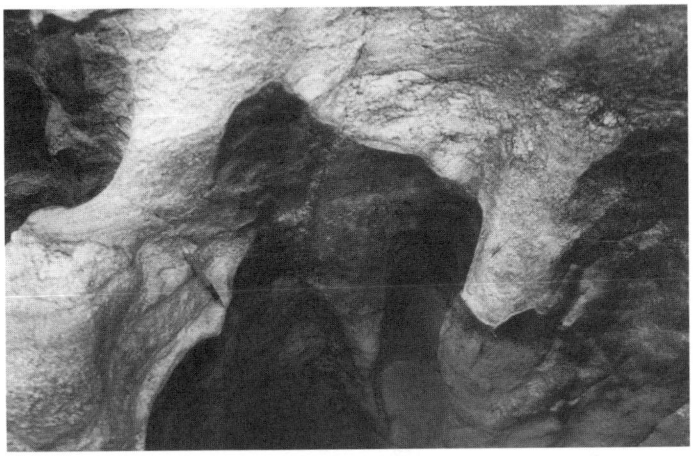

Figure 162. *Pothole* with passive sharp rim arising through surface planation (of the light-colored area around the *pothole*). Nam Mae Chaem (Ob Luang), Thailand. Granitic gneiss. Pen for scale.

that which generated them (i.e., their removal by planation). It requires that the *potholes* are no longer active. The rough planation surface in the example cited suggests that it is occurring mainly by subaerial, rather than fluvial, erosion.

- Through breaching. A *pothole*, when first breached, either by plucking or by the gradual erosion and growth of the *pothole*, will probably have two passive relatively sharp, near-vertical rims where the thickness of the wall tapers to zero (Fig. 161B [ii]). However, one of these will face into the flow and will experience rapid erosion and rounding (Fig. 21). The other, downstream-facing rim may remain sharp, depending on the flow structure that develops within and around the *pothole* subsequent to its breaching. Such a downstream-facing rim may become an active and therefore permanent high angle sharp edge, formed where recirculating water exiting the *breached pothole* is confluent with flow in the main channel, and where both bodies of water separate from the boundary. However, many *lateral potholes* have rounded rims on all sides, and therefore their hydrodynamic environment and internal flow structures are not conducive to the development of sharp edges. It must also be possible, therefore, for a *breached pothole* to be formed in such an environment and to develop such an internal flow pattern, and therefore for both vertical rims of the *breached pothole* to become rounded.

Mean Flow Pathlines Associated with Active Sharp Edges

Allen (1971a) produced detailed skin friction maps for *flutes*, which have cuspate rims on their upstream sides, based on experiments using plaster of Paris. Skin friction lines, which are mean flow pathlines immediately adjacent to the boundary, may be determined from very small-scale sculpting within bedforms. His results are summarized in Figures 160A, 160D, and 163A. *Flutes* contain areas of upstream-directed and downstream-directed skin friction lines (i.e., they contain separated flow), divided by lines of reattachment. Flow separates from the boundary at the rim and at any internal ridges (such as lateral and *median ridges*) within the *flute*, and reattaches at concave areas of the floor of the *flute*. As a result of the parabolic shape of the *flute* rim, however, skin friction lines in general converge on the rim in an oblique rather than perpendicular fashion. Allen (1971a) used these maps to deduce the existence of attached vortices and associated helical mean flow pathlines within *flutes*. Only in a small region at the very proximal part of the *flute* rim, where the rim is orientated transverse to the flow, are skin friction lines perpendicular to the rim. Only in these areas are the skin friction lines directly rather than obliquely convergent; the mean flow pathlines therefore are rotational but not helical. In the case of very weakly convergent (i.e., highly oblique) skin friction lines, the ridges (such as *median ridges*) are rounded rather than sharp, and the mean flow paths are essentially parallel and streamwise, and only weakly vortical. In the case of *spindle-shaped flutes*, there are no upstream-directed skin friction lines; instead, flow separates into a pair of counterrotating longitudinal vortices within the *flute*.

It may be noted that Kor et al. (1991), Shaw (1996), and Bryant (2001: p. 95) attempted to apply the results of Allen (1971a) to infer the nature of the flow structures associated with the s-forms sichelwannen, muschelbrüche, and open spindle flutes. Their proposed flow structures are unrealistic, however, because they involve a single vortex rather than a counterrotating pair within spindle flutes, and a straight rather than a horseshoe vortex within muschelbrüche. These flow patterns result in flow reattachment rather than flow separation at some sharp edges. In the case of sichelwannen, a single horseshoe-shaped vortex is proposed, rather than the double horseshoe vortex proposed by Allen (1971a) for the analogous *flute with median ridge* (Fig. 163A).

However, Allen (1971a) did not produce skin friction maps for conjugate *flutes* or mature assemblages of *scallops*. In order to rectify this, a flume study on a plaster of Paris replica of naturally occurring *scallops* was undertaken. A polyurethane foam mold (Fig. 164) was made from a *scallop* assemblage at Sleightholme Beck, and this mold was used to cast the plaster of Paris replica. The replica was then placed in a flume and subjected to a flow calculated to be appropriate to the measured spacing of the *scallops* according to the empirical relationship between flow velocity and *scallop* wavelength given by Goodchild and Ford (1971). Small-scale sculpting developed associated with bubbles in the plaster of Paris, and skin friction lines as indicated by the small-scale sculpting were mapped from a photomosaic of the model. Part of the skin friction map is shown in Figure 163B. In common with the case of (isolate) *flutes*, there are both upstream- and downstream-directed skin friction lines, and therefore flow separates at the *scallop* crests. However, in contrast with *flutes*, skin friction lines have a tendency to approach crests in a perpendicular fashion, indicating a predominance of roller-type attached vortices and circular rather than helical mean

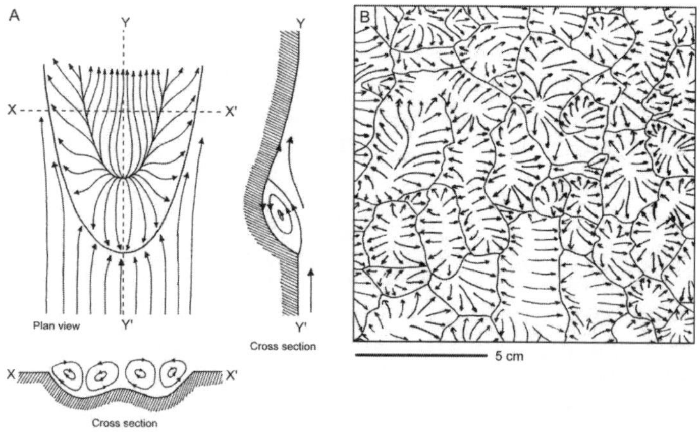

Figure 163. Flow patterns associated with *flutes* and *scallops*. A: Idealized flow structures in and around a *flute with median ridge* (modified after Allen, 1971a). A *flute* without a *median ridge* would contain a single rather than a double horseshoe vortex. B: Skin friction lines in a *directional scallop* assemblage, obtained experimentally in plaster of Paris by the authors. Current left to right.

Figure 164. Polyurethane mold taken of *scallop* assemblage at Sleightholme Beck, UK. Direction of flow was approximately top to bottom. Lighting is from the bottom of the picture. Note the resemblance to turbidite sole structures. The marks on the scale are in centimeters.

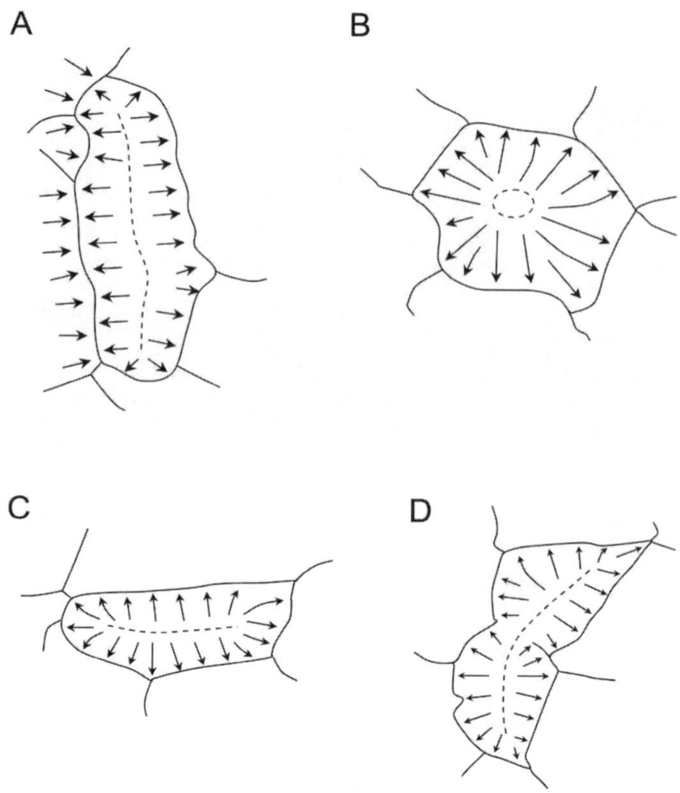

Figure 165. Idealized patterns of skin friction lines in various morphologies of *directional scallops*, as defined by their length:width ratios. A: Transverse axial (L < W). B: Radial (L ≈ W). C: Longitudinal axial (L > W). D: Composite. Dashed lines denote lines or regions of flow separation. Flow from left to right.

flow pathlines. This observation illustrates a tendency for boundary layer reattachment near the center of each *scallop*, with skin friction lines radiating from the region of reattachment toward the crests. The region of reattachment may be linear, leading to an axial pattern of skin friction line divergence, or it may be a small, apparently near-circular region, leading to a radial pattern of skin friction lines within the *scallop*. This difference in skin friction lines between isolate *flutes* and *scallops* illustrates how flow patterns within isolate individuals differ from those within conjugate assemblages, and the effect of interactions between neighboring individuals in such assemblages. Four types of skin friction patterns may be identified in *scallops*, based on the streamwise length L and the width W of the host depression (Fig. 165):

(i) Axial (L < W). In *scallops*, which are wider that they are long, the region of reattachment is a line transverse to the flow, indicating a roller-type attached vortex in the lee of the upstream crest.

(ii) Radial (L ≈ W). In equidimensional *scallops*, the region of reattachment often approximates a small circular area. The attached vortices in such *scallops* are thought to resemble truncated versions of the horseshoe-shaped vortices present in isolate *flutes*, but with mean flow pathlines which are less helical that those in flutes.

(iii) Axial (L > W). *Scallops* that are longer than they are wide are relatively rare, but where they do occur, the region of reattachment is a line orientated approximately in a streamwise direction, indicating quasi-streamwise, counterrotating vortex pairs.

(iv) Composite. In more complex *scallops*, two distinct regions with different patterns of skin friction lines may be observed, either an axial-radial combination or two axial regions of different orientation. Such *scallops* probably either are in the process of subdividing or are two merging *scallops*. The destruction (i.e., merging) of *scallops* and the generation of new ones within an assemblage are both documented processes (Allen, 1971a; Blumberg and Curl, 1974).

Further Comments on the Interpretation of Bedforms in Open Bedrock Channels

Bedforms as Indicators of Flow Patterns

The above discussion relating to sharp edges and their associated flow patterns suggests that the potential exists for the development of a set of indicators to be used in the qualitative interpretation of flow structures within and around individual bedrock bedforms and within systems of contiguous forms. It should always be remembered that sculpted forms can only ever be guides to local flow patterns within the immediate vicinity of the bedform and at the range of channel discharges over which the bedform is active. However, once the hydrodynamics of such individuals and systems are understood, the nature and distribution of sculpted forms may be used to reconstruct channel-scale flow patterns that exist at a range of stages, including high stage when direct observation may be impractical. These indicators include the following (illustrated schematically in Figure 166), but this list is not expected to be comprehensive.

(i) Fine-scale sculpting within bedforms, i.e., *lineations* and *spindle-shaped flutes* (Fig. 166A), where present, can be used to produce skin friction maps of bedforms (Maxson, 1940). This technique was applied by Allen (1968, 1971a) to plaster of Paris models to determine the flow structures present within ripples and *flutes* respectively (e.g., Fig. 163A). In this technique, skin friction lines are indicated by the sculpted "tails" developed in the wake of small defects in the substrate. These defects and their tails are much smaller than the bedform within which they reside, and they do not themselves grow into larger sculpted forms because they are below the critical defect size for continued growth (Allen, 1971a). The authors also used this technique to determine flow patterns associated with *scallops* (Fig. 163B). Figure 167A shows a natural example of fine-scale sculpting. At a larger scale, *scallops* themselves do sometimes adorn the walls of larger bedforms such as *potholes* and can be used to infer recirculating flow (e.g., Springer and Wohl, 2002) (Fig. 167B).

(ii) As discussed above in relation to *flutes* and *scallops*, at the crests and rims of sculpted forms flow separates from the boundary, and skin friction lines are convergent, while in concave areas flow reattaches to the boundary, and skin friction lines are divergent (Figs. 160 and 163). In the case of isolate *flutes* (Fig. 163A), skin friction lines are in general oblique to the rim and internal ridges, indicating helical flow within attached vortices. In the case of *scallops* (Fig. 163B), skin friction lines are in general perpendicular to crests, indicating radial and axial patterns of flow separation. These results allow an analysis of the flow structures present in any bedform to be made by an interpretation of the distribution of convex and concave surfaces, especially the distribution of sharp edges, and by an analysis of the effects of any contiguous neighbors. Crests and rims trending in a transverse direction must generate roller-type vortices and skin friction lines approximately perpendicular to the crest. Depending on the morphol-

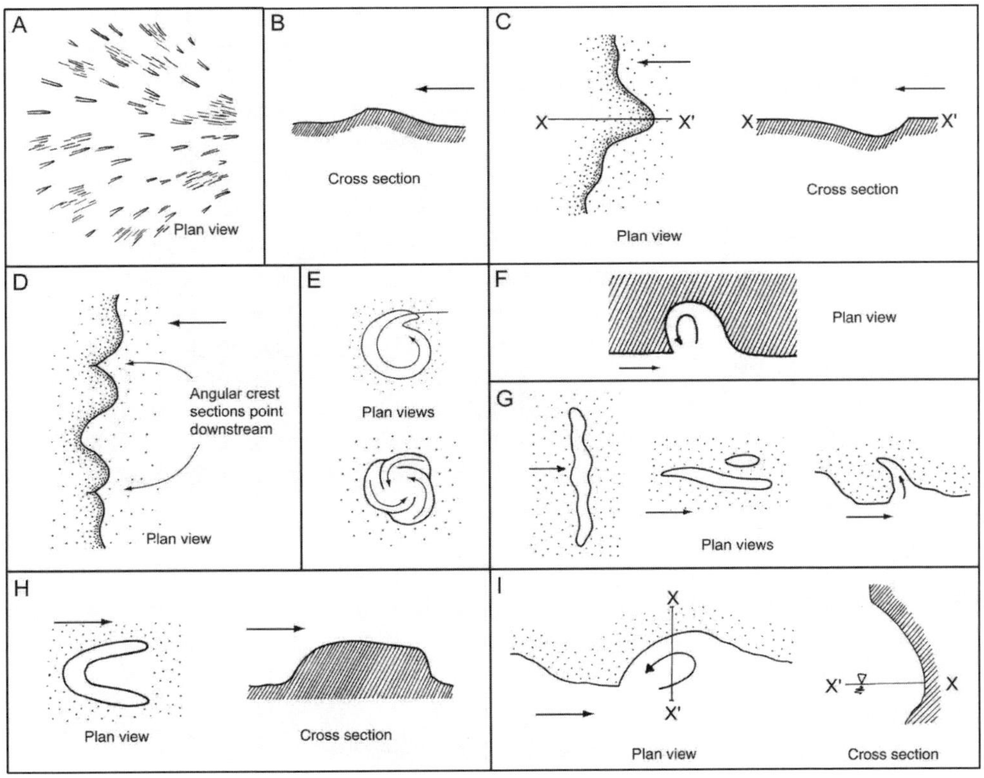

Figure 166. Bedforms useful as indicators of local flow direction (shown by arrows). A: Fine sculpting indicative of a divergent flow pattern from right to left. B: Asymmetrical sharp edge. C: Parabolic scour hollow. D: Angular sections of sharp crests. E: *Pothole with entry furrow* (top) and *pothole with spiral furrows* (bottom). F: *Lateral pothole* with sharp rim on the upstream side. G: *Transverse furrow* (left), *longitudinal furrows* (center), *reversed furrow* (right). H: *Obstacle mark* (left) and *upstream-facing convex surface* (right). I: *Overhanging concave surface.*

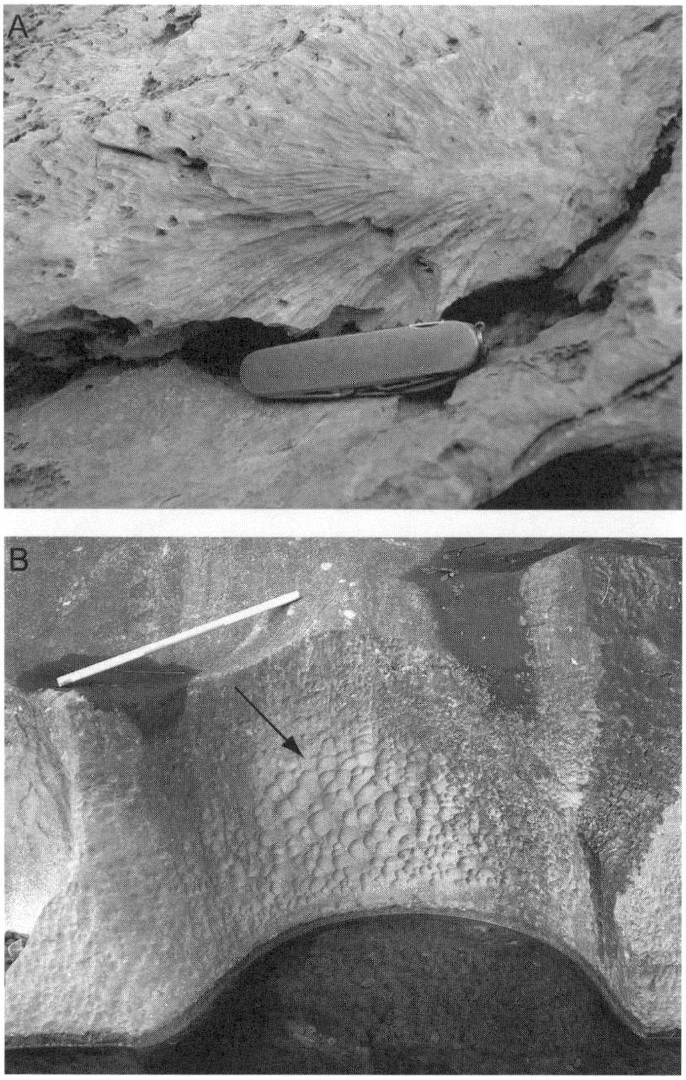

Figure 167. A: Fine-scale sculpting on a bedrock surface indicating a divergent pattern of mean flow vectors. Nahal Neqarot, Israel. Flow is from right to left. Limestone. Penknife for scale. B: *Scallops* (arrowed) indicating the presence of separated flow in a *breached pothole*. The *scallops* indicate flow from left to right, but flow in the main channel is in the opposite direction. Ease Gill, UK. Limestone. The scale is 60 cm long.

ogy of any associated contiguous *concave features*, crests at other orientations will generate vortices containing more or less helical mean flow pathlines and more or less obliquely convergent skin friction lines. For example, the sharp longitudinal crests (*bladed forms*) in Figure 94 are interpreted as indicating skin friction lines obliquely convergent on those crests, which represent streamwise flow separation lines with counter-rotating vortices and helical mean flow pathlines on either side.

(iii) Low-angle sharp edges that are asymmetrical, with convex stoss sides and steeper, concave lee faces, are clear indicators of the mean local flow direction (Fig. 166B). Examples include *SCHF* (*sharp-crested hummocky forms*) and *directional scallops*. The lee faces of asymmetrical edges may also be darkened with algal growth, while the stoss sides are cleaner.

(iv) High-angle sharp edges point downstream, while gently rounded and blunted convex surfaces face upstream. For example, Figure 168 shows a complex bedform that is interpreted as a *pothole* with two *entry furrows* and an *exit furrow*. Local flow is from the top left of the photograph. The two *entry furrows* (labeled En) meet the *pothole* tangentially, they have an orientation approximately transverse to the local flow direction, and their upstream edges are overhanging high-angle edges pointing downstream. The *exit furrow* (labeled Ex) is broad and leaves the *pothole* radially on the downstream side. Also on the downstream side of the *pothole* are blunt and gently rounded convex surfaces (labeled B) facing upstream. Jennings (1983) also describes how high-angle sharp edges indicate flow direction, in this case in the form of *transverse furrows* that overhang on the upstream side.

(v) Scoured hollows with arcuate, parabolic (in plan view) upstream crests within a pattern of *scallops* or *SCHF* are analogous to *flutes*, and the crests are convex in the (local) upstream direction (Fig. 166C). Isolated transverse sharp

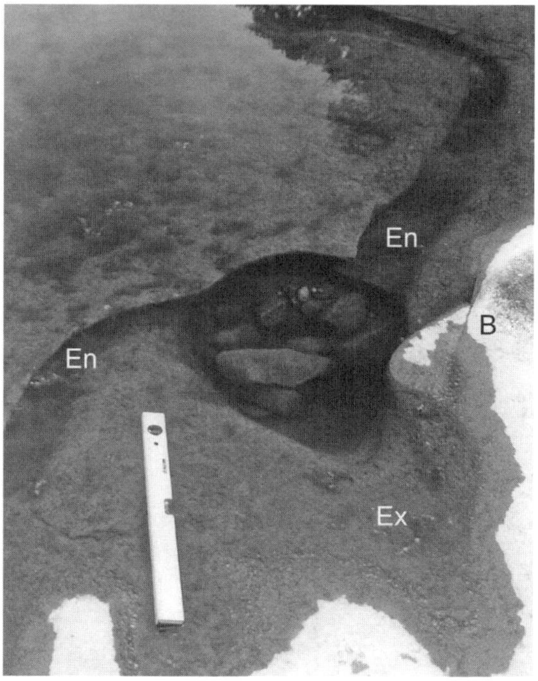

Figure 168. *Pothole* with two *entry furrows* (En) and an *exit furrow* (Ex). River Lune (Halton), UK. Flow from top left to bottom right. Fine-grained sandstone. The scale is 60 cm long. Several features of this bedform are aids to interpreting local flow direction, including the position of the *exit furrow*, the blunted and rounded surfaces (B), which face upstream, and the overhanging upstream rim of the bedform, which forms a high-angle sharp edge pointing downstream. The *pothole* is partially filled with cobbles.

edges may also contain similar arcuate parabolic crest sections enclosing scoured hollows.

(vi) Transverse sharp edges often possess angular sections (seen in plan view), which point in the local downstream direction (Fig. 166D). Examples include *directional scallops*, *streamlined steep lee faces*, and *SCHF*.

(vii) Active *potholes*, whether complete, *lateral*, or *breached*, may reasonably be assumed to contain vertically orientated vortices, and their sense of rotation may be interpreted with the assistance of *entry furrows*, *spiral furrows*, *yin-yang furrow complexes*, and sharp vertical rims (Figs. 160 and 166E). In the case of *hierarchical* and *coalesced potholes*, it is sometimes found that the distribution of sharp and rounded vertical rims between component *potholes* may be used to reconstruct internally consistent systems of counterrotating, concordant vortices.

(viii) In the case of *lateral potholes* that possess one sharp vertical rim and one rounded vertical rim, the sharp rim must be on the (locally) upstream side (Fig. 166F). As the flow within the *pothole* must also be concordant with flow in the main channel as it separates from the sharp edge, this also gives the sense of rotation of the vortex within the *pothole*.

(ix) *Longitudinal*, *transverse*, and *reversed furrows* are normally readily distinguished on morphological grounds and by interpretation of the flow structures induced by the topography of their immediate environment, as noted in the descriptions given in the typology, and can thus be used as indicators of mean local flow vectors (Fig. 166G).

(x) *Obstacle marks* and *upstream-facing convex surfaces* are readily interpreted and are good indicators of local mean flow directions (Fig. 166H).

(xi) *Overhanging concave surfaces* are good indicators of recirculating gyres with diameters on the meter scale (Fig. 166I). *Reversed furrows* are associated with separated flow in reentrants and in the lee of projections, are typically curved and tapering, and are also good indicators of recirculating gyres.

Bedforms as Indicators of the Relative Roles of Bedload and Suspended Load in Erosion

Is abrasion by bedload an important factor in the generation of sculpted forms? Foley (1980b) and Sklar and Dietrich (1998) use models that assume that most, if not all, channel incision is effected through bedload erosion. In contrast, Zen and Prestegaard (1994) and Whipple et al. (2000a) use morphological evidence to infer that erosion by suspended load is dominant in the bedrock channels they describe. In this section, several lines of evidence are advanced that point to the greater importance in general of the role of suspended load over that of bedload in channels that are not structurally controlled, i.e., those channels where plucking is not the dominant mechanism of incision.

(i) The distribution of bedforms. It was noted above that sculpted forms are not restricted to either the lowest or the horizontal portions of the channel boundary, but that active bedforms may also adorn vertical surfaces and relatively raised portions of the channel, where abrasion by bedload must be either absent or ineffective.

(ii) The role of "grinders" in *potholes*. In the classical model of *pothole* formation, erosion is achieved by the rotary movement of pebbles ("grinders") trapped within the *pothole*. Alexander (1932), however, in experiments on the hydraulics of *potholes*, observed that the amount of energy supplied to tools at the bottom of the hole decreases rapidly with depth, and that the spiral currents within the holes were capable of removing only very fine particles. He concluded that in *potholes* whose depth exceeded significantly their diameter, large grinders would in fact be a hindrance to erosion because they would tend to pack and fill the hole, and that *potholes* could only form in water with very small amounts of bedload. Generally, Alexander (1932) favored sand-grade sediment as the "tools" responsible for pothole erosion. Whipple et al. (2000a) use the principle of "surface capture efficiency" to show that the complex and intricate internal morphology of some *potholes* is inconsistent with the grinder hypothesis (see below). A further difficulty for the grinder hypothesis noted by Alexander (1932) is that in excavations of *potholes*, supposed grinders are found to be packed along with sand and clay in a fabric-supported manner, which argues against the rocks ever being active grinders. Also, grinders cannot be instrumental in the formation of *lateral potholes*, since they cannot be retained in the *pothole*.

(iii) Morphological evidence (I): the scale of bedforms. Whipple et al. (2000a) discuss the principle of the "capture efficiency" of a surface, which is the idea that in order to impact with and erode a surface, transported particles must become decoupled from the flow and travel under the influence of their own momentum to some extent. Depending upon their size, particles become decoupled at different radii of curvature of the flow pathlines; large particles deviate from flow pathlines even at low curvature, while very fine particles follow flow pathlines faithfully. The implications of this principle are that small-scale bedforms, and small-scale, intricate morphology within larger ones, are indicative of abrasion by suspended load because they indicate a high degree of coupling between fluid flow and erosion patterns. Of the sites studied for this report, in those that are hydrodynamically controlled, bedforms on the scale of centimeters to decimeters are the rule. In many cases, such features completely cover the channel boundary. It is further observed that erosional forms that attain dimensions of several decimeters to meters become host to small-scale secondary sculpting within them. Lorenc et al. (1994), for example, describe secondary sculpting developing within large *potholes*, and this phenomenon also occurs within *flutes*, *short furrows*, and *obstacle marks*. Such erosion patterns cannot be produced by bedload; on the contrary, bedload erosion would tend to destroy heavily ornamented boundaries on account of the high momentum of the clasts and their low degree of coupling to the flow. Zen and Prestegaard (1994) demonstrate the importance of abrasion by suspended load with their description of erosion within *lateral potholes* that was sensitive to variation in rock strength on the millimeter scale, producing a ribbed effect. A further consideration is that the erosive particles, be they bedload or suspended load, must be much smaller than the smallest sculpted forms for which they are responsible (with the obvious exception of *tool marks*). In some cases, sculpted forms are similar in size to, or are smaller than, most bedload clasts (Tinkler, 1993), which again rules out bedload as the erosive agent for such features.

(iv) Morphological evidence (II): erosion by separated flows. Many *concave features* either occur on surfaces within flow separation cells, such as on the lee sides of obstructions, or do themselves constitute a flow separation cell. In order for the eroding particles to reach such areas, they must be highly coupled to the flow. For example, within *flutes*, the highest rate of erosion occurs close to the point of reattachment (Allen, 1971a), which requires that the sediment particles responsible for abrading the bed be tightly coupled to the flow pathlines and follow them faithfully into the *flute*, i.e., it requires the particles to be suspended. Allen (1971a) considers that the presence of bedload actually destroys *flutes*. Hancock et al. (1998) and Whipple et al. (2000a) describe intense erosion due to *potholes* and *flutes* on the lee sides of boulders, and clearly within flow separation cells. In fact, these authors use the distribution of bedforms on boulders and rock projections in the River Ganges, which tend to have smooth upstream faces and heavily potholed and fluted lee faces (section 3.3), and the existence of *overhanging flutes* (section 1.2b), to infer that erosion is most intense on the lee faces of boulder and defects, that in their words, "attack is from the back." Although this does indeed appear to be the case, a cautionary note should be added that in reaching this conclusion an important fact has been overlooked. For while deep potholing and fluting on the lee of obstacles to the flow is an impressive and obvious sign of erosion by suspended load, and the volume of the sculpted forms gives a minimum estimate of the amount of rock removed, a completely unquantifiable amount of rock has been removed from the upstream side, where bedload erosion may have been significant.

(v) Morphological evidence (III): the presence of sharp crests and rims. Bedload displays little coupling to flow pathlines. Erosion by bedload will therefore be focused on the upstream sides of obstacles to the flow, and it will tend to produce linear features and smooth surfaces with large radii of curvature and without small-scale sculpting. This is because any raised areas will receive harder and more frequent impacts, and therefore with time their topography will become subdued, and because erosion cannot be effected by bedload within small depressions. In particular, delicate ornamentation such as sharp crests and rims would disappear. However, the prevalence of positive features with sharp transverse crests in bedrock channels has been noted above, and *concave features* with cuspate rims are not uncommon. This indicates that erosion by bedload is unimportant in the great number of bedrock channels in which these features abound. Furthermore, sharp edges may be considered to be indicative not only of the existence of flow separation, but by extension, also of the dominance of the erosion associated with flow separation, i.e., abrasion by suspended load, because if the sharp edges are stable features, the erosion they induce must act to sustain them.

(vi) Morphological evidence (IV): the convexity of upstream-facing surfaces. Sharp (1949) describes the development of upwind-facing boulder surfaces subject to erosion by saltating sand grains in aeolian environments. Such surfaces may be convex, planar, or concave. Indeed, convex and concave surfaces are considered by Sharp (1949) as stages in the development of the planar surfaces of mature faceted boulders. On upstream-facing surfaces in fluvial environments, however, planar surfaces are rare, and in the experience of the authors, there is a complete lack of concave surfaces. Upstream-facing surfaces are overwhelmingly convex, and usually strongly so. This suggests that the erosive mechanism in fluvial environments is different from that active in the development of faceted boulders in aeolian environments, and therefore that suspended rather than saltating grains are the agents for it.

(vii) Erosion by dissolution. In the highly successful experimental studies of *flutes* of Allen (1971a), it was an explicit assumption that the dissolution of a rough bed of plaster of Paris in a turbulent flow is directly analogous to the erosion of a mud bed by abrasion by suspended load under similar circumstances. The bedforms he generated were of a small scale with intricate morphology and cuspate rims, and the observed associated mean flow structures within the *flutes* involve high streamline curvature. This confirms that the analogy is a correct one, and that fluid flow and erosion patterns in *solutional forms* are also tightly coupled. An exception to this is *deep circular solution pits*, whose large depth relative to their diameter would inhibit such coupling. Instead, these forms appear to depend on a more passive process of dissolution in relatively stagnant water within the *pit*. In the case of all types of *solution pits*, given their scale, their sometimes intricate morphology, and/or the presence of sharp crests, abrasion by bedload is clearly not significant.

(viii) Transcritical flow. In bedrock channels, critical and transcritical flow is often observed to be maintained in portions of the channel over a wide range of stages (Tinkler, 1993,

1997b; Tinkler and Wohl, 1998a). Henderson (1966: p. 29–57) shows that critical flow minimizes specific energy for a given discharge and is therefore the most energy-efficient flow regime. Grant (1997) suggested that critical flow is both a limiting condition in fluvial hydraulics and a state toward which river channels tend by mutual adjustment of the flow with the boundary, even in bedrock channels. The importance of standing waves and hydraulic jumps (i.e., transcritical flow) in locally enhancing erosion by suspended load has already been discussed. It is the experience of the authors that at low stages, undular flow and hydraulic jumps are often maintained within *runnels*, even in very low flows when most of the channel is dry. These *runnels* sometimes contain sculpted forms orientated appropriately to flow within the *runnel* at low stage, but not to flow parallel to the main channel, which occurs at higher stages. This suggests that transcritical flow is important in locally enhancing erosion by suspended load over a wide range of discharges in bedrock channels, and that this erosion is effective even at very low stages within portions of the channel.

(ix) The rarity of tool marks. If abrasion by bedload was significant, one would expect to find an abundance of tool marks of all kinds, whereas in fact they are very rare in the reaches examined for this study. Whipple et al. (2000a) also found this to be the case, although percussion marks are abundant in some channels (N. Hovius, 2001, personal commun.).

In passing, it may be noted that some of these criteria may also be used to make inferences as to the mechanical properties of the eroding fluid. For instance, Gjessing (1967) postulated that *potholes* formed under glaciers were the result of scouring by a viscous or plastic material such as a saturated ground moraine or a mixture of water, ice, and rock particles. Given the intricate internal morphology of some of the *potholes* he studied, however, including delicate sharp edges, *internal short furrows*, and *spiral furrows*, and the implications of these in terms of the presence of small-scale flow separation cells and high streamline curvature, this hypothesis is clearly incorrect.

The Relative Importance of Erosion by Suspended Load and Bedload: A Hypothesis

The interpretation of bedforms in this manner should be exercised with caution because, whereas the products of abrasion by suspended load (such as most of the *concave features*) are often highly visible, abrasion by bedload would tend to subdue, rather than enhance, relief at short wavelengths, and its effects are therefore often less visible. The only evidence of abrasion by bedload is a lack of small-scale, intricate and delicate sculpture. Such negative evidence is rarely convincing, or particularly noticeable. For example, erosion by bedload remains a possibility for relatively large-wavelength, low-relief, gently rounded bedforms such as (smooth-crested) *hummocky forms* and *partially abraded surfaces*. Other relatively large-scale features lacking detailed ornamentation and occupying a low position within the channel, such as some *runnels*, *cavettos*, *inner channels*, and *upstream-facing convex surfaces* also remain candidates for erosion by bedload. K. Tinkler (2000, personal commun.) reports an abraded *runnel* from Cooksville Creek, Ontario, which he suggests is eroded by bedload. Also, bedload abrasion may be more important in some reaches or within certain parts of a channel cross section. For instance, in some bedrock channels at certain times, the authors have observed narrow, freshly abraded tracks that are lighter in color than surrounding bedrock and that are clear of algal growth. These are evidence of locally enhanced erosion, and given their simple, linear form and position occupying the lowest parts of the channel, erosion by bedload would appear to be the mechanism responsible for them. Furthermore, these tracks have been found to be conduits of enhanced bedload transport, relative to adjacent parts of the channel floor (Richardson et al., 2003). It should also be borne in mind when observing sculpted forms in bedrock channels, and particularly when making generalizations, that except in ephemeral streams, the lowest portions of the channel are usually inaccessible.

Nevertheless, collectively the observations in the preceding section suggest the general dominance of erosion by suspended load over that by bedload in hydrodynamically controlled channels in which there are abundant sculpted forms. Three possible conclusions may be drawn from this inferred dominance:

(i) Bedload transport rates are insignificant in bedrock channels.

(ii) Bedload transport rates are generally significant, but bedload is vastly outcompeted in terms of erosion by suspended load.

(iii) Bedload transport is significant, but tends to be routed and concentrated along relatively narrow pathways such as the freshly abraded tracks described above and is effective only in these areas.

The first of these possibilities seems unlikely. The existence of a bedrock channel does not imply a low supply of sediment to the channel, but rather that the sediment supply is exceeded by the transport capacity of the channel in the long term (Howard et al., 1994). Furthermore, bedrock channels rarely have boundaries composed purely of bedrock, and in fact generally contain bars of sediment or alternating gravel-bed and bedrock reaches, or have alluvial reaches both upstream and downstream of a bedrock reach (see, e.g., Keller and Melhorn, 1978; Ashley et al., 1988; Miller, 1991; Wohl, 1992a, 1992b, 1993; Wohl et al., 1994; Gupta et al., 1999; Wende, 1999; Kasai et al., 2004). If such rivers are in steady state, this implies that bedrock channels are not channels of exceptionally low sediment supply but are simply zones of enhanced sediment transport capacity through which sediment is conveyed rapidly between bars and alluvial reaches where transport capacity is lower and sediment storage is more permanent. The authors do not wish to imply, however, that there are no bedrock channels that are truly sediment starved, and in which bedload transport is insignificant. The second alternative, likewise, does not seem particularly feasible; where bedload transport rates are significant, so ought to be their effects. The possibility of suspended load sculpting sharp edges and fine-scale, intricate morphology—and these not being rounded off by bedload impacts—does not seem a likely one.

Therefore, it is hypothesized that although bedload transport rates in bedrock channels are generally significant, the clasts travel along narrowly defined preferred pathways within the channel, such as the freshly abraded tracks observed by the authors in some channels, and therefore the effects of bedload transport are restricted to these areas. These pathways may not be in exposed parts of the channel, even at very low flow, and therefore remain invisible. Elsewhere, erosion by suspended load is dominant (in hydrodynamically controlled channels), and sculpted forms are developed with a range of morphologies and at a range of scales.

The Influence of Substrate on Bedform Development in Open Bedrock Channels

Bedforms in Carbonate Rocks

Limestones (and calcareous siliciclastic rocks to a lesser degree) develop the most diverse arrays of bedforms and the densest accumulations of them. They also exhibit the best-formed examples of bedforms in general. It would appear therefore that dissolution in limestone can in general proceed more rapidly than abrasion in silicate rocks.

The universality of the majority of bedforms across different lithologies refutes suggestions that those in limestone are exhumed karst features. However, limestone and other highly calcareous rocks do develop a suite of endemic bedforms, the *solutional forms*. These are not restricted to limestone but are typical of limestone and are poorly developed elsewhere. In particular, the various types of *solution pits* and *scallops* are characteristic of limestone, and these are bedforms that occur on a scale much smaller than most sculpted forms in other lithologies. They are therefore indicative of formation by dissolution. Also, while the rounded types of *hummocky forms* are not uncommon in other lithologies, they are especially common and well developed in limestone.

Bedforms in Silicate Rocks

In general, lithology per se exerts a relatively small influence on the style of sculpted forms in bedrock channels, as long as the rock is not very coarse-grained or heterogeneous. There does not appear to be an association between bedform type and rock type. There is, however, a strong dependence of the style of bedform development on structural aspects of the substrate. For the purposes of this report, rock structure can be considered as any feature of the rock that imparts anisotropy or heterogeneity to the substrate at a scale at which it affects the variation in erosion rates across the channel boundary. This is a broad class of features that includes (1) the rock fabric and differences between individual grains and between grains and matrix in very coarse-grained rocks; (2) primary internal structures arising through the mode of formation of the rock such as bedding, sedimentary structures, and fossils in sedimentary rocks, and flow banding, chilled margins, and crystal settling in igneous rocks; (3) diagenetic effects such as differential compaction and cementation; (4) metamorphic structures such as cleavage, schistosity, and gneissicity; (5) structures arising through brittle deformation, such as joints, brittle shear zones, veins, and dikes; (6) heterogeneity in grain size, composition, or strength arising through any other mechanism.

In rocks with such structures, bedforms are often structurally influenced but may still be abundant. The structural influence may be manifested in the distribution of bedforms, in the orientation of bedforms, or in the shape of bedforms, or any combination of these. This influence arises through the enhancement of erosion in weaker parts of the rock, or its enhancement in directions in which the rock is weak, for example, *furrows* carved out along bedding planes. Rocks with pronounced and/or closely spaced structures show limited or absent bedform development, and the channel morphology becomes structurally controlled. In some cases, this is because the rock structure itself, and not the flow patterns in the water, determine the variation in abrasion rates over the surface, leading to *irregular surfaces* and chaotic flow patterns. However, in cases where the rock contains closely spaced fractures, or structures along which fracture readily occurs, it is because plucking or flaking become the dominant erosion mechanism (Whipple et al., 2000a). For instance, shale erodes mainly by flaking during the wetting and drying cycle, and by displacement of small blocks through freeze-thaw and hydraulic plucking (Tinkler and Parish, 1998). These processes, not abrasion, are the rate-limiting ones in such rocks.

However, the influence of rock structures on bedform development is not always destructive; bedforms sometimes originate at defects, structurally controlled topographic features, and heterogeneities. The most obvious examples are *flutes*. *Potholes* and some *parallel-sided furrows* can also be seen to originate at such sites, presumably through vortices pinned to and shed from the defects. Furthermore, larger projections and obstacles to the flow, many of which have a structural origin, generate larger flow separation cells that can lead to the development of features such as *reversed furrows*, *transverse furrows*, *lateral potholes*, *obstacle marks*, and *overhanging concave surfaces*.

The smallest scale of sculpted forms that can develop in a bedrock channel, and the "quality" of the forms, are determined by the rock texture and hardness. Hard, fine-grained rocks are capable of developing the smallest features. Bedforms in such rocks are also better defined and more diverse. Soft and coarse-grained rocks are not capable of supporting fine sculptured detail, either because the grain size becomes comparable with that of the bedforms and dominates erosion patterns or because the rock tends to crumble. Similarly, rock structure is sometimes important in determining the scale of the smallest bedforms, for similar reasons as grain size.

The Scale of Bedforms in Open Bedrock Channels

There are six variables, not all of them independent, each of which seems capable of acting as a control on the scale of sculpted forms under appropriate circumstances, and these are described below. The first five variables apply to trains or assemblages of similar forms in channels developing either uniform flow or regularly varying flow, and they control the scale of individuals within equilibrium assemblages. The final variable is rather different and applies to isolate individuals in any channel.

(i) Channel width. *Pool-riffle* sequences in bedrock are known to have a wavelength of five to seven times the channel width, in common with their alluvial counterparts (Keller and Melhorn, 1978). Furthermore, all large-scale bedforms (section 6 of the typology) exist on a scale comparable with that of the channel width; their maximum lateral extent is therefore that of the channel.

(ii) Flow depth. *Groups of parallel-sided furrows* exist in channels in mud, whose lateral spacing is found to be approximately twice the bankfull depth of the channel (Ikeda, 1978). The observed spacing was attributed to streamwise vortices occupying the full flow depth. A similar control may exist for examples of *groups of parallel-sided furrows* in bedrock, which are regularly spaced and extensive, and which exist on *concordant plane-bed* channels with uniform flow (e.g., King, 1927; Blank, 1958; Wohl, 1993).

(iii) Bedform Reynolds number and flow velocity. Curl (1966) showed through dimensional analysis that there exists a Reynolds number based on *scallop* wavelength and flow velocity that is for all practical purposes a constant. This results in the inverse relationship between the wavelength of mature *scallops* and near-bed flow velocity discussed in a preceding section. The same principle should apply to any periodic bedform involving mass transfer from the boundary to the flow in which molecular diffusion plays a negligible role (Curl, 1966; Allen, 1971a). Indeed, ripples on the undersides of ice covers have been shown to exhibit a similar relationship between wavelength and flow velocity (Ashton and Kennedy, 1972; Epstein and Cheung, 1983).

(iv) Froude number. It has been suggested that *step-pool* systems in gravel-bed streams form as antidunes during extreme floods, which then degrade into *step-pool* morphology, and that *step-pools* form in approximately critical flow (e.g., Whittaker and Jaeggi, 1982; Chartrand and Whiting, 2000). If bedrock channels are considered to be deformable on the time scale of extreme discharges (Wohl, 1992a), then this mechanism must also be considered as a possibility for the origin of *step-pools* in bedrock channels.

(v) Channel slope. Allen (1983) proposed that transverse ribs in gravel-bed channels form under hydraulic jumps and derived an expression for the minimum spacing of ribs, assuming that a hydraulic jump, and therefore a rib, cannot occur until the flow has traveled a sufficient distance from the previous rib to accelerate back to the critical state. This expression predicts that for a given size of grains in the ribs, the minimum spacing is inversely proportional to channel slope. In this expression, grain size is used pragmatically as a surrogate for critical flow depth. If *SCHF* (*sharp-crested hummocky forms*) within *runnels* are also formed under hydraulic jumps, as discussed above, then a relationship similar to that proposed by Allen (1983) for rib spacing should exist between the spacing of *SCHF* and *runnel* slope, provided that a suitable measure of critical flow depth is found.

(vi) The scale of locally generated turbulent coherent structures. While channel width and flow depth (at some formative discharge) can clearly act as controls on the scale of bedforms in uniform flow in *plane beds*, it is likely that this is not commonly the case, because of the more usual irregularity of bedrock channel boundaries, and because of the dominance of nonuniform flow. These features of bedrock channels act to generate complex flow structures, such as recirculating gyres of various scales and hydraulic jumps, and to prevent the formation of a fully developed boundary layer so that the boundary in general is not subject to a uniform flow depth or channel width. The majority of bedforms described in this typology have characteristic dimensions that are a small fraction of the formative flow depth, and they are not responding to, or generating, eddies that occupy the full flow depth. Many bedforms, such as *transverse furrows*, *reversed furrows*, *obstacle marks*, those *parallel-sided furrows* that are directly related to topographical "defects," and the early stages of *potholes* and *flutes* represent the boundary responding passively to the turbulent coherent structures imposed on it by local topography. As such bedforms grow, however, the turbulent coherent structures that the bedforms themselves generate become increasingly important. It is possible that some bedforms remain passive and attain and remain at a size that merely reflects the scale of turbulent structures imposed upon them by one or more roughness elements in their immediate environment. In many bedforms, however, turbulent structures generated internally supersede the effects of the original defect, thus controlling the future development of the bedform. In this way, the boundary is no longer acting passively but rather is actively controlling its own development. In mature *flutes*, for instance, the original defect is destroyed, and its shape has no influence on the shape of the *flute* beyond a certain age. The *flute* becomes self-sustaining, and its development is independent of the nature of the original defect (Allen, 1971a). Other bedforms, such as the majority of *short furrows* and *parallel-sided furrows*, are not clearly related to turbulent structures generated by local topography and are also likely to be actively controlling their own development in this way. It should be noted that the above discussion applies only to isolate individuals. Once bedforms become interfering, an assemblage develops in which both external influences (i.e., interactions between individuals) and internal influences are important to individual depressions. All isolate bedforms, except those that have become moribund, will enlarge with time. It is observed, however, that large bedforms tend to become subject to secondary sculpting, as for example in the case of *compound potholes* and large *flutes* and *obstacle marks*. It thus appears that in a given environment, there is a maximum stable size of individuals beyond which the persistent internal flow structures within bedforms tend to break up into smaller eddies, generating secondary sculpting on a smaller scale, within the large primary bedform. In this manner, simple morphologies evolve into compound ones.

CONCLUSIONS

Bedforms in bedrock channels are defined as sculpted forms produced by mechanisms other than plucking, except in very large bedforms where plucked blocks are much smaller that the bedform. The influence of the structure of the substrate on erosion processes and channel morphology, or the absence of that influence, has been used to classify open bedrock channels into structurally controlled, structurally influenced, and hydrodynamically controlled types. In the first type of channel, erosion by plucking is dominant, and there is a lack of sculpted forms. In the last type of channel, erosion by abrasion or dissolution is dominant and the boundary is sculpted into a variety of bedforms dictated by hydrodynamic flow patterns. In structurally influenced channels, bedforms are present but they are influenced by structural elements of the substrate. A consistent terminology has been developed to describe the morphology of bedforms and their relationships with each other. Bedforms in bedrock channels are analogous to their sedimentary counterparts in some respects, especially in their hydrodynamic control, but also differ fundamentally in several ways. These include the higher resistance of the substrate to erosion and the consequent slow rate of evolution of bedforms, the unidirectionality of the erosion process, and the fact that in general, gravitational forces are unimportant and therefore bedrock bedforms are not restricted to the floor of the channel. Bedrock channels differ fundamentally from their alluvial counterparts in having very rough and relatively undeformable boundaries, and in developing highly nonuniform and complex flow.

Bedforms in bedrock channels exhibit great diversity. They have been classified primarily using morphology, but also according to their hydrodynamic environment and by the processes responsible for them, where appropriate. The major division of bedforms is into *concave features*, *convex* and *undulating surfaces*, *composite forms*, *solutional forms*, and *tool marks*. *Composite forms* comprise elements from both *concave features* and *convex* and *undulating surfaces*. A final group consists of large-scale features that have dimensions comparable to the width of the channel and which act as topography rather than as channel roughness. Within these groups, approximately 140 identifiable bedform types are described in detail. In addition, field examples of structures generated in laboratory dissolution studies by Allen (1971a) are described for the first time.

Four major principles apply to the development of bedforms in bedrock channels. It is observed that there is continuity of form between end-member types, by which one bedform grades into another by means of many subtle variations. Thus, most bedforms represent end members of or midpoints within a spectrum of forms. Bedforms also exhibit convergence of form, by which analogous morphologies arise in different settings and via different processes. Three different routes by which bedform morphology may converge are identified. Thirdly, it is observed that when bedforms are contiguous and interact, they do so in such a manner that their associated flow structures interfere in a mutually reinforcing and constructive manner. Lastly, the observed almost ubiquitous nature of sharp-crested transverse forms of one type or another suggests that these represent a stable state to which bedrock surfaces naturally tend, where they are able to do so.

The origins of two types of sharp-crested transverse features are discussed. The defect theory of the origin of *scallops* is supported, with the description of a new additional mechanism for their development involving the generation of patterns of *space-filling solution pits*. *Scallop* formation is seen as a two-step process involving, firstly, an assemblage of *flutes* or *solution pits* becoming fully conjugate, and secondly, a period of evolution during which the wavelength of the *scallops* becomes adjusted to the flow and in which their planform evolves from three-dimensional toward a more two-dimensional one. Several mechanisms are proposed for the origin of *sharp-crested hummocky forms*, all of which are thought to be operative, either separately or together, under appropriate circumstances. The phenomenon of sharp edges in bedforms in general is discussed. Sharp edges are classified into two types: active and passive. Active sharp edges are those which represent lines of flow separation and discontinuities in erosion rate profiles. It is proposed that the presence of strongly convergent skin friction lines is both a necessary and sufficient condition for their development. Once developed, active sharp edges influence the flow in such a way that they become enhanced by erosion. Active sharp edges can be subdivided into low-angle and high-angle types. At low-angle sharp edges, a boundary layer separates, forming a flow separation cell in which flow converges discordantly with flow in the boundary layer approaching the edge. At high-angle sharp edges, the flow approaching the edge from either side converges concordantly at the edge. Passive sharp edges are those which are incidental to the process through which they arise and do not develop inevitably through that process. Once developed, they play no active role in their evolution.

The interpretation of bedforms in bedrock channels is discussed further in relation to the understanding of flow patterns within individual bedforms and within entire channel reaches. A set of guidelines is developed by which such patterns can be inferred. Furthermore, the study of the morphology and distribution of bedforms indicates that abrasion by bedload is generally insignificant compared to that achieved by suspended load in channels that are hydrodynamically, rather than structurally, controlled. It is hypothesized that this is because, although bedload transport rates in bedrock channels are not insignificant, the bedload is routed along narrow pathways within the channel, and therefore erosion by bedload is restricted to these areas.

Although lithology per se does not exert a great influence on the development of sculpted forms, structural elements of the substrate do. This influence can be reflected in the shape, orientation, and distribution of bedforms, or in the absence of bedforms resulting from the complete structural control of channel morphology and erosion. However, it is observed that limestone develops the most diverse, the most densely packed, and the best-developed accumulations of bedforms. Also, limestone develops a suite of endemic features, indicative of formation by corrosion, which are rare or absent in silicate rocks. Six parameters, some of them interdependent, are thought to be potential controls on the scale of bedforms, of which the most important is the scale of locally generated turbulent coherent structures.

ACKNOWLEDGMENTS

The authors would like to thank Hiroshi Ikeda of the University of Tsukuba, Japan, for his interesting discussions in the field, and Kosit Lorsirirat of the Royal Irrigation Department, Thailand, for his practical assistance. Noam Greenbaum, Tamir Grodek, and Itai Haviv of the Ben Gurion University of the Negev and the Hebrew University of Jerusalem provided assistance with fieldwork in Israel. We would particularly like to thank all these people for their hospitality. The British Council provided funding for fieldwork in Thailand and Japan. The British Geomorphological Research Group assisted with funding to present this typology at the 5th International Conference in Geomorphology, Tokyo, 2001. Andy Russell is thanked for the photographs of bedforms in Greenland, and John Shaw for the pictures of cavitation marks.

REFERENCES CITED

Acarlar, M.S., and Smith, C.R., 1987, A study of hairpin vortices in a laminar boundary layer, part 1: Hairpin vortices generated by a hemispherical protuberance: Journal of Fluid Mechanics, v. 175, p. 1–41.

Alexander, H.S., 1932, Pothole erosion: Journal of Geology, v. 40, p. 305–337.

Allen, J.R.L., 1965, Scour marks in snow: Journal of Sedimentary Petrology, v. 35, p. 331–338.

Allen, J.R.L., 1968, Current ripples: Their relation to patterns of water and sediment motion: Amsterdam, North Holland Publishing Company, 433 p.

Allen, J.R.L., 1969, Erosional current marks of weakly cohesive mud beds: Journal of Sedimentary Petrology, v. 39, p. 607–623.

Allen, J.R.L., 1971a, Transverse erosional marks of mud and rock: Their physical basis and geological significance: Sedimentary Geology, v. 5, p. 167–385.

Allen, J.R.L., 1971b, Bed forms due to mass transfer in turbulent flows: A kaleidoscope of phenomena: Journal of Fluid Mechanics, v. 49, p. 49–63.

Allen, J.R.L., 1974, Sedimentology of the Old Red Sandstone (Siluro-Devonian) in the Clee Hills area, Shropshire, England: Sedimentary Geology, v. 12, p. 73–167.

Allen, J.R.L., 1982, Sedimentary structures: Their character and physical basis, Volume 2: Amsterdam, Elsevier, Developments in Sedimentology, v. 30B, 663 p.

Allen, J.R.L., 1983, A simplified cascade model for transverse stone-ribs in gravelly streams: Royal Society of London Proceedings, ser. A, v. 385, p. 253–266.

Allen, J.R.L., 1987, Streamwise erosional structures in muddy sediments of the Severn Estuary, southwestern England: Geografisker Annaler, ser. A, v. 69, p. 37–46.

Ängeby, O., 1951, Pothole erosion in recent waterfalls: Lund Studies in Geography, ser. A, Physical Geography, No. 2, p. 1–34.

Arndt, R.E.A., 1981, Cavitation in fluid machinery and hydraulic structures: Annual Reviews of Fluid Mechanics, v. 13, p. 273–328.

Ashley, G.M., Renwick, W.H., and Haag, G.H., 1988, Channel form and process in bedrock and alluvial reaches of the Raritan River, New Jersey: Geology, v. 16, p. 436–439, doi: 10.1130/0091-7613(1988)016<0436:CFAPIB>2.3.CO;2.

Ashton, G.D., and Kennedy, J.F., 1972, Ripples on the underside of river ice covers: American Society of Civil Engineers Proceedings: Journal of the Hydraulics Division, v. 98, p. 1603–1624.

Baker, V.R., 1973, Erosional forms and processes for the catastrophic Pleistocene Missoula floods in eastern Washington, *in* Morisawa, M., ed., Fluvial geomorphology: Binghamton, New York, State University of New York, a proceedings volume of the 4th annual geomorphology symposia series held at Binghamton, New York, 1973, p. 123–148.

Baker, V.R., 1978, Large scale erosional and depositional features of the Chaneled Scabland, *in* Baker, V.R., and Nummedal, D., eds., The Channeled Scabland: A guide to the geomorphology of the Columbia Basin, Washington: Washington, D.C., NASA, p. 81–115.

Baker, V.R., and Kale, V.S., 1998, The role of extreme floods in shaping bedrock channels, *in* Tinkler, K.J., and Wohl, E.E., eds., Rivers over rock: Fluvial processes in bedrock channels: Washington, D.C., American Geophysical Union, Geophysical Monograph 107, 323 p.

Baker, V.R., and Pickup, G., 1987, Flood geomorphology of the Katherine Gorge, Northern Territory, Australia: Geological Society of America Bulletin, v. 98, p. 635–646, doi: 10.1130/0016-7606(1987)98<635:FGOTKG>2.0.CO;2.

Best, J., 1996, The fluid dynamics of small-scale alluvial bedforms, *in* Carling, P.A., and Dawson, M.R., eds., Advances in fluvial dynamics and stratigraphy: Chichester, John Wiley and Sons, 530 p.

Blank, H.R., 1958, Pothole grooves in the bed of the James River, Mason County, Texas: Texas Journal of Science, v. 10, p. 292–301.

Blumberg, P.N., and Curl, R.L., 1974, Experimental and theoretical studies of dissolution roughness: Journal of Fluid Mechanics, v. 5, p. 735–742.

Bögli, A., 1981, Solution of limestone and karren formation, *in* Sweeting, M.M., ed., Karst geomorphology: Stroudsberg, Pennsylvania, Hutchinson Ross, Benchmark Paper in Geology 59, 427 p.

Bradwell, T., 2005, Bedrock megagrooves in Assynt, NW Scotland: Geomorphology, v. 65, p. 195–204.

Braley, S.M., and Wilson, P., 1997, Ventifacts from the coast of Northumberland: Geologists' Association Proceedings, v. 108, p. 7.

Bretz, J.H., 1924, The Dalles type of river channel: Journal of Geology, v. 32, p. 139–149.

Bryant, E., 2001, Tsunami: The underrated hazard: New York, Cambridge University Press, 326 p.

Carling, P.A., and Tinkler, K., 1998, Conditions for the entrainment of cuboid boulders in bedrock streams: An historical review of literature with respect to recent investigations, *in* Tinkler, K.J., and Wohl, E.E., eds., Rivers over rock: Fluvial processes in bedrock channels: Washington, D.C., American Geophysical Union, Geophysical Monograph 107, 323 p.

Chartrand, S.M., and Whiting, P.J., 2000, Alluvial architecture in headwater streams with special emphasis on step-pool topography: Earth Surface Processes and Landforms, v. 25, p. 583–600, doi: 10.1002/1096-9837(200006)25:63.0.CO;2-3.

Coleman, J.C., 1949, An indicator of water-flow in caves: University of Bristol Spelaeological Society Proceedings, v. 6, p. 57–67.

Collinson, J.D., and Thompson, D.B., 1982, Sedimentary structures: London, Allen and Unwin, 194 p.

Curl, R.L., 1966, Scallops and flutes: Cave Research Group of Great Britain Transactions, v. 7, p. 121–160.

Curl, R.L., 1974, Deducing flow velocity in cave conduits from scallops: National Speleological Society Bulletin, v. 36, p. 1–5.

Defina, A., 1996, Transverse spacing of low-speed streaks in a channel flow over a rough bed, *in* Ashworth, P.J., et al., eds., Coherent flow structures in open channels: Chichester, Wiley and Sons, 733 p.

De Wit, M.C.J., 1999, Post-Gondwana drainage and the development of diamond placers in western South Africa: Economic Geology and the Bulletin of the Society of Economic Geologists, v. 94, p. 721–740.

Diffendal, J.F., 1982, Gully, scour hole and pothole development at the base of the Gering Formation (Miocene?), southeastern Banner County, Nebraska: Contributions to Geology, University of Wyoming, v. 21, p. 1–6.

Dixon, E.E.L., 1921, The geology of the South Wales Coalfield, Part XIII: The country around Pembroke and Tenby: London, Her Majesty's Stationery Office, Memoirs of the Geological Survey of Great Britain, 220 p.

Duckson, D.W., and Duckson, L.J., 1995, Morphology of bedrock step pool systems: Water Resources Bulletin, v. 31, p. 43–51.

Dzulynski, S., and Sanders, J.E., 1962, Current marks on firm mud bottoms: Transactions of the Connecticut Academy of Arts and Sciences, v. 42, p. 57–96.

Dzulynski, S., and Walton, E.K., 1965, Sedimentary features of flysch and greywackes: London, Elsevier, 274 p.

Elston, E.D., 1917, Potholes: Their variety, origin and significance (I): Scientific Monthly, v. 5, p. 554–567.

Elston, E.D., 1918, Potholes: Their variety, origin and significance (II): Scientific Monthly, v. 6, p. 37–51.

Epstein, M., and Cheung, F.B., 1983, Complex freezing-melting interfaces in fluid flow: Annual Review of Fluid Mechanics, v. 15, p. 293–319.

Fairbridge, R.W., editor, 1968, The encyclopedia of geomorphology: Stroudsberg, Pennsylvania, Dowden, Hutchinson and Ross, Encyclopedia of Earth Science Series, Vol. 3, 1295 p.

Foley, M.G., 1980a, Quaternary diversion and incision, Dearborn River, Montana: Geological Society of America Bulletin, v. 91, p. 2152–2188.

Foley, M.G., 1980b, Bedrock incision by streams: Geological Society of America Bulletin, v. 91, p. 2189–2213.

Ford, D., and Williams, R., 1991, Karst geomorphology and hydrology: London, Chapman and Hall, 601 p.

Fralick, P., 1999, Paleohydraulics of chute-and-pool structures in a Paleoproterozoic fluvial sandstone: Sedimentary Geology, v. 125, p. 129–134.

Friend, P.F., 1965, Fluviatile sedimentary structures in the Wood Bay series (Devonian) of Spitsbergen: Sedimentology, v. 5, p. 39–68.

Gale, S.J., 1984, The hydraulics of conduit flow in carbonate aquifers: Journal of Hydrology, v. 70, p. 309–327, doi: 10.1016/0022-1694(84)90129-X.

Gardner, T.W., 1983, Experimental study of knickpoint and longitudinal profile evolution in cohesive, homogeneous material: Geological Society of America Bulletin, v. 94, p. 664–672, doi: 10.1130/0016-7606(1983)94<664:ESOKAL>2.0.CO;2.

Gilbert, G.K., 1906, Moulin work under glaciers: Geological Society of America Bulletin, v. 17, p. 317–320.

Ginés, A., 1996, An environmental approach to the typology of karren landform assemblages in a Mediterranean mid-mountain karst: The Serra de Tramuntana, Mallorca, Spain, *in* Formós, J.-J., and Ginés, A., eds., Karren landforms: Palma de Mallorca, Universitat de les Isles Balears, 450 p.

Gjessing, J., 1967, Potholes in connection with plastic scouring forms: Geografisker Annaler, ser. A, v. 49, p. 178–187.

Glasser, N.F., and Nicholson, F.H., 1998, Subglacial meltwater erosion at Loch Treig: Scottish Journal of Geology, v. 34, p. 7–13.

Goodchild, M.F., and Ford, D.C., 1971, Analysis of scallop patterns by simulation under controlled conditions: Journal of Geology, v. 79, p. 52–62.

Grant, G.E., 1997, Critical flow constrains flow hydraulics in mobile-bed streams: A new hypothesis: Water Resources Research, v. 33, p. 349–358, doi: 10.1029/96WR03134.

Grass, A.J., and Mansour-Tehrani, M., 1996, Generalised scaling of coherent bursting structures in the near-wall region of turbulent flow over smooth and rough boundaries, *in* Ashworth, P.J., et al., eds., Coherent flow structures in open channels: Chichester, Wiley and Sons, 733 p.

Grass, A.J., Stuart, R.J., and Mansour-Tehrani, M., 1991, Vortical structures and coherent motion in turbulent flow over rough boundaries: Royal Society of London Philosophical Transactions, ser. A, v. 336, p. 35–65.

Gupta, A., Kale, V.S., and Rajaguru, S.N., 1999, The Narmada river, India, through space and time, *in* Miller, A.J., and Gupta, A., eds., Varieties of fluvial form: Chichester, Wiley and Sons, 521 p.

Hancock, G.S., Anderson, R.S., and Whipple, K.X., 1998, Beyond power: Bedrock river incision process and form, *in* Tinkler, K.J., and Wohl, E.E., eds., Rivers over rock: Fluvial processes in bedrock channels: Washington, D.C., American Geophysical Union, Geophysical Monograph 107, 323 p.

Henderson, F.M., 1966, Open channel flow: New York, Macmillan, 522 p.

Heritage, G.L., van Niekerk, A.W., and Moon, B.P., 1999, Geomorphology of the Sabie River, South Africa: An incised bedrock-influenced channel, *in* Miller, A.J., and Gupta, A., eds., Varieties of fluvial form: Chichester, Wiley and Sons, 521 p.

Howard, A.D., 1998, Long profile development of bedrock channels: Interaction of weathering, mass wasting, bed erosion and sediment transport, *in* Tinkler, K.J., and Wohl, E.E., eds., Rivers over rock: Fluvial processes in bedrock channels: Washington, D.C., American Geophysical Union, Geophysical Monograph 107, 323 p.

Howard, A.D., Seidl, M.A., and Dietrich, W.E., 1994, Modeling fluvial erosion on regional to continental scales: Journal of Geophysical Research, v. 99, p. 13,971–13,986.

Hoyt, J.H., and Henry, V.J., 1963, Rhomboid ripple mark, indicator of current direction and environment: Journal of Sedimentary Petrology, v. 33, p. 604–608.

Hutchinson, D.W., 1996, Runnels, rinnenkarren and mäanderkarren: Form, classification and relationships, *in* Formós, J.-J., and Ginés, A., eds., Karren landforms: Palma de Mallorca, Universitat de les Isles Balears, 450 p.

Ikeda, H., 1978, Large-scale grooves formed by scour on cohesive mud surfaces: Bulletin of the Environmental Research Center, University of Tsukuba, Japan, v. 2, p. 91–95.

Itakura, M., and Ikeda, H., 1997, On the channel configuration of plane-bed bedrock channels on the upper Obitsu River basin: Bulletin of the Environmental Research Center, University of Tsukuba, Japan, v. 22, p. 9–21.

Ives, R.L., 1948, Plunge pools, potholes and related features: Rocks and Minerals, v. 23, p. 3–10.

Izumi, N., and Parker, G., 2000, Linear stability analysis of channel inception: Downstream- driven theory: Journal of Fluid Mechanics, v. 419, p. 239–262, doi: 10.1017/S0022112000001427.

Jacob, R.J., Bluck, B.J., and Ward, J.D., 1999, Tertiary-age diamondiferous fluvial deposits of the lower Orange River valley, southwestern Africa: Economic Geology and the Bulletin of the Society of Economic Geologists, v. 94, p. 749–758.

Jennings, J.N., 1983, Swirlholes and related bedrock river channel forms: Australian Geographer, v. 15, p. 411–414.

Jennings, J.N., 1985, Karst geomorphology: Oxford, Blackwell, 293 p.

Johnsson, G., 1988, Potholes: Glacial and non-glacial cavities: Geografisker Annaler, ser. A, v. 70A, p. 333–336.

Kale, V.S., Baker, V.R., and Mishra, S., 1996, Multi-channel patterns of bedrock rivers: An example from the central Narmada basin, India: Catena, v. 26, p. 85–98, doi: 10.1016/0341-8162(95)00035-6.

Kale, V.S., and Joshi, V.U., 2004, Evidence of formation of potholes in bedrock on human timescale: Indrayani River, Pune District, Maharashtra: Current Science, v. 86, p. 723–726.

Kale, V.S., and Shingade, B.S., 1987, A morphological study of potholes of Indrayani Knickpoint (Maharashtra), *in* Datya, V.S., et al., eds., Explorations in the Tropics: Pune, India, Professor K.R. Dikshit Felicitation Committee, Professor K.R. Dikshit Felicitation Volume, p. 206–214.

Karcz, I., 1967, Harrow marks: Current-aligned, sedimentary structures: Journal of Geology, v. 75, p. 113–120.

Karcz, I., 1968, Fluviatile obstacle marks from the wadis of the Negev (southern Israel): Journal of Sedimentary Petrology, v. 38, p. 1000–1012.

Karcz, I., 1973, Reflections on the origin of source small-scale longitudinal streambed scours, *in* Morisawa, M., ed., Fluvial geomorphology: Binghamton, New York, State University of New York, a proceedings volume of the 4th annual geomorphology symposia series held at Binghamton, New York, 1973, p. 149–174.

Kasai, M., Marutani, T., and Brierley, G., 2004, Channel bed adjustments following major aggradation in a steep headwater setting: Geomorphology, v. 62, p. 199–214.

Keller, E.A., and Melhorn, W.N., 1978, Rhythmic spacing and origin of pools and riffles: Geological Society of America Bulletin, v. 89, p. 723–730, doi: 10.1130/0016-7606(1978)89<723:RSAOOP>2.0.CO;2.

King, P.B., 1927, Corrosion and corrasion on Barton Creek, Austin, Texas: Journal of Geology, v. 35, p. 631–638.

Kline, S.J.W., Reynolds, W.C., and Schraub, F.A., 1967, The structure of turbulent boundary layers: Journal of Fluid Mechanics, v. 30, p. 741–773.

Knighton, D., 1998, Fluvial forms and processes: New York, Wiley and Sons, 383 p.

Kobor, J.S., and Roering, J.J., 2004, Systematic variation of bedrock channel gradients in the central Oregon Coast Range: Implications for rock uplift and shallow landsliding: Geomorphology, v. 62, p. 239–256, doi: 10.1016/j.geomorph.2004.02.013.

Kodama, Y., and Ikeda, H., 1984, An experimental study on the formation of Yakama Potholes, *in* Research Report of "Yakama Potholes", Special Natural Commemoration: Yanadani, Ehime Prefecture, Japan, Yakama Research Association, 187 p. (in Japanese).

Kor, P.S.G., and Cowell, D.W., 1998, Evidence for catastrophic subglacial meltwater sheetflood events on the Bruce Peninsula, Ontario: Canadian Journal of. Earth Sciences, v. 35, p. 1180–1202.

Kor, P.S.G., Shaw, J.A., and Sharpe, D.R., 1991, Erosion of bedrock by subglacial meltwater, Georgia Bay, Ontario—A regional view: Canadian Journal of Earth Sciences, v. 28, p. 623–642.

Koyama, T., and Ikeda, H., 1998, Effect of riverbed gradient on bedrock channel configuration: A flume experiment: Proceedings of the Environmental Research Center, University of Tsukuba, Japan, v. 23, p. 25–34.

Kunert, M., and Coniglio, M., 2002, Origin of vertical shafts in bedrock along the Eramosa River valley near Guelph, southern Ontario: Canadian Journal of Earth Sciences, v. 39, p. 43–52, doi: 10.1139/e01-053.

Lauritzen, S.E., Ive, A., and Wilkinson, B., 1983, Mean annual runoff and the scallop flow regime in a subarctic environment: British Cave Research Association Transactions, v. 10, p. 97–102.

Leeder, M.R., 1982, Sedimentology: London, Chapman and Hall, 344 p.

Livingstone, I., and Warren, A., 1996, Aeolian geomorphology: Harlow, UK, Longman, 211 p.

Lorenc, M.W., Barco, P.M., and Saavedra, J., 1994, The evolution of potholes in granite bedrock, western Spain: Catena, v. 22, p. 265–274, doi: 10.1016/0341-8162(94)90037-X.

Lugn, A.L., 1939, Classification of the Tertiary System in Nebraska: Geological Society of America Bulletin, v. 50, p. 1245–1276.

Lugt, H.J., 1983, Vortex flow in nature and technology: New York, Wiley and Sons, 297 p.

Maxson, J.H., 1940, Fluting and faceting of rock fragments: Journal of Geology, v. 48, p. 717–751.

Maxson, J.H., and Campbell, I., 1935, Stream fluting and stream erosion: Journal of Geology, v. 43, p. 729–744.

Miller, J.R., 1991, The influence of bedrock geology on knickpoint development and channel bed configuration along downcutting streams in south-central Indiana: Journal of Geology, v. 99, p. 591–605.

Montgomery, D.R., and Gran, K.B., 2001, Downstream variations in the width of bedrock channels: Water Resources Research, v. 37, p. 1841–1846.

Morgan, A.V., 1970, Late Weichselian potholes near Wolverhampton, England: Journal of Glaciology, v. 9, p. 125–133.

Murphy, P.J., Hall, A.M., and Cordingley, J.N., 2000, Anomalous scallop distributions in Joint Hole, Chapel-le-Dale, North Yorkshire, UK: Cave and Karst Science, v. 27, p. 29–32.

Nemec, W., Lorenc, M.W., and Alonso, J.S., 1982, Potholed granite terrace in the Rio Salor valley, western Spain: A study of bedrock erosion by floods: Tecniterrae, v. 50, p. 6–21.

Otvos, E.G., 1965, Types of rhomboid beach surface patterns: American Journal of Science, v. 263, p. 271–276.

Parker, G., and Izumi, N., 2000, Purely erosional cyclic and solitary steps created by flow over a cohesive bed: Journal of Fluid Mechanics, v. 419, p. 203–238, doi: 10.1017/S0022112000001403.

Peabody, F.E., 1947, Current crescents in the Triassic Moenkopi Formation: Journal of Sedimentary Petrology, v. 17, p. 73–76.

Pohn, H.A., 1983, The relationship of joints and stream drainage in flat-lying rocks of south-central New York and northern Pennsylvania: Zeitschrift für Geomorphologie N.F., v. 27, p. 375–384.

Pollard, A., Wakarani, N., and Shaw, J., 1996, Genesis and morphology of erosional shapes associated with turbulent flow over a forward-facing step, *in* Ashworth, P.J., et al., eds., Coherent flow structures in open channels: Chichester, Wiley and Sons, 733 p.

Reusser, L.J., Bierman, P.R., Pavich, M.J., Zen, E-an, Larsen, J., and Finkel, R., 2004, Rapid Late Pleistocene incision of Atlantic passive-margin river gorges: Science, v. 305, p. 499–502.

Richardson, K., Benson, I., and Carling, P.A., 2003, An instrument to record sediment movement in bedrock channels, *in* Bogen, J., et al., eds., Erosion and sediment transport in rivers: Technological and methodological advances: Wallingford, UK, International Association of Hydrological Sciences, Publication no. 283 (proceedings of the IAHS Oslo workshop, June 2002), 238 p.

Richardson, P.D., 1968, The generation of scour marks near obstacles: Journal of Sedimentary Petrology, v. 38, p. 965–970.

Righter, K., 1997, High bedrock incision rates in the Atanguillo River valley, Jalisco, western Mexico: Earth Surface Processes and Landforms, v. 22, p. 337–344, doi: 10.1002/(SICI)1096-9837(199704)22:43.3.CO;2-T.

Sato, S., Matsuura, H., and Mayazaki, M., 1987, Potholes in Shikoku, Japan, Part 1: Potholes and their hydrodynamics in the Kurokawa River, Ehime: Memoirs of the Faculty of Education of Ehime University, Natural Science, v. 7, p. 127–190.

Schlyter, P., 1994, Paleo-periglacial ventifact formation by suspended silt or snow: Site studies in south Sweden: Geografiska Annaler, ser. A, v. 76, p. 15.

Schumm, S.A., 1991, To interpret the Earth: Ten ways to be wrong: Cambridge, UK, Cambridge University Press, 133 p.

Seidl, M.A., Weissel, J.K., and Pratson, L.F., 1996, Kinematics of escarpment retreat across the rifted continental margin of S.E. Australia: Basin Research, v. 8, p. 301–316, doi: 10.1046/j.1365-2117.1996.00266.x.

Selby, M.J., 1977, Transverse erosional marks on ventifacts from Antarctica: New Zealand Journal of Geology and Geophysics, v. 20, p. 949–969.

Selby, M.J., 1985, Earth's changing surface: Oxford, UK, Oxford University Press, 607 p.

Sevon, W.D., 1988, Exotically sculpted diabase: Pennsylvania Geology, v. 20, p. 2–7.

Sharp, R.P., 1949, Pleistocene ventifacts east of the Big Horn Mountains, Wyoming: Journal of Geology, v. 57, p. 175–195, 3 plates.

Shaw, J., 1996, A meltwater model for Laurentide subglacial landscapes, *in* McCann, S.B., and Ford, D.C., eds., Geomorphology sans frontières: Chichester, UK, John Wiley and Sons, p. 181–236.

Shaw, J., Faragini, D.M., Kvill, D.R., and Rains, R.B., 2000, The Athabasca fluting field, Alberta, Canada: Implications for the formation of large-scale fluting (erosional lineations): Quaternary Science Reviews, v. 19, p. 959–980.

Shepherd, R.G., and Schumm, S.A., 1974, Experimental study of river incision: Geological Society of America Bulletin, v. 85, p. 257–268, doi: 10.1130/0016-7606(1974)85<257:ESORI>2.0.CO;2.

Shrock, R.R., 1948, Sequence in layered rocks: New York, McGraw-Hill, 507 p.

Sklar, L., and Dietrich, W.E., 1998, River longitudinal profiles and bedrock incision models: Streampower and the influence of sediment supply, *in* Tinkler, K.J., and Wohl, E.E., eds., Rivers over rock: Fluvial processes in bedrock channels: Washington, D.C., American Geophysical Union, Geophysical Monograph 107, 323 p.

Slabe, T., 1996, Cave rocky relief and its speleogenetical significance: Ljubljana, Znanstvenoraziskovalni Center SAZU (Slovenian Academy of Sciences and Arts), Zbirka ZRC (Scientific Research Center) publication no. 10, 128 p.

Smith, D.I., Greenaway, M.A., Moses, C., and Spate, A.P., 1995, Limestone weathering in eastern Australia, Part 1: erosion rates: Earth Surface Processes and Landforms, v. 20, p. 451–463.

Spaggiari, R.I., Ward, J.D., and de Wit, M.C.J., 1999, Fluvial characteristics of the diamondiferous Droogeveldt Gravels, Vaal Valley, South Africa: Economic Geology and the Bulletin of the Society of Economic Geologists, v. 94, p. 741–748.

Sparks, B.W., 1972, Geomorphology (2nd edition): London, Longman, 530 p.

Springer, G.S., and Wohl, E.E., 2002, Empirical and theoretical investigations of sculpted forms in Buckeye Creek, west Virginia: Journal of Geology, v. 110, p. 469–481.

Summerfield, M.A., 1991, Global geomorphology: Harlow, UK, Longman, 537 p.

Suzuki, M., and Ikeda, H., 1994, Origin of flat-floored bedrock channel in the Chiiwa River, a tributary of the Upper Toyo River, Aichi Prefecture: Bulletin of the Environmental Research Center, University of Tsukuba, Japan, v. 19, p. 45–56 (in Japanese).

Sweeting, M.M., 1972, Karst landforms: London, Macmillan, 362 p.

Thybony, S., 2000, Playing the slots: National Geographic, v. 198, no. 1, p. 118–131.

Tinkler, K.J., 1993, Fluvially sculpted rock bedforms in Twenty Mile Creek, Niagara Peninsula, Ontario: Canadian Journal of Earth Sciences, v. 30, p. 945–953.

Tinkler, K.J., 1997a, Rockbed wear at a flow convergence zone in Fifteen Mile Creek, Niagara Peninsula, Ontario: Journal of Geology, v. 105, p. 263–274.

Tinkler, K.J., 1997b, Critical flow in rockbed streams with estimated values of Manning's *n*: Geomorphology, v. 20, p. 147–164.

Tinkler, K.J., and Parish, J., 1998, Recent adjustments to the long profile of Cooksville Creek, an urbanized bedrock channel in Mississauga,

Ontario, *in* Tinkler, K.J., and Wohl, E.E., eds., Rivers over rock: Fluvial processes in bedrock channels: Washington, D.C., American Geophysical Union, Geophysical Monograph 107, 323 p.

Tinkler, K.J., and Stenson, R.E., 1992, Sculpted bedrock forms along the Niagara Escarpment, Niagara Peninsula, Ontario: Géographie Physique et Quaternaire, v. 46, p. 195–207.

Tinkler, K.J., and Wohl, E.E., 1998a, A primer on bedrock channels, *in* Tinkler, K.J., and Wohl, E.E., eds., Rivers over rock: Fluvial processes in bedrock channels: Washington, D.C., American Geophysical Union, Geophysical Monograph 107, 323 p.

Tinkler, K.J., and Wohl, E.E., 1998b, Field studies of bedrock channels, *in* Tinkler, K.J., and Wohl, E.E., eds., Rivers over rock: Fluvial processes in bedrock channels: Washington, D.C., American Geophysical Union, Geophysical Monograph 107, 323 p.

Tooth, S., and McCarthy, T.S., 2004, Anabranching in mixed bedrock-alluvial rivers: The example of the Orange River above the Augrabies Falls, Northern Cape Province, South Africa: Geomorphology, v. 57, p. 235–262, doi: 10.1016/S0169-555X(03)00105-3.

Tritton, D.J., 1988, Physical fluid dynamics (2nd edition): Oxford, UK, Clarendon Press, 519 p.

Trudgill, S.T., 1985, Limestone geomorphology: Harlow, UK, Longman, Geomorphology Text No. 8, 196 p.

van Niekerk, A.W., Heritage, G.L., Broadhurst, L.J., and Moon, B.P., 1999, Bedrock anabranching channel systems: Morphology and dynamics in the Sabie River, Mpumalanga Province, South Africa, *in* Miller, A.J., and Gupta, A., eds., Varieties of fluvial form: Chichester, Wiley and Sons, 521 p.

Weissel, J.K., and Seidl, M.A., 1997, Influence of rock strength properties on escarpment retreat across passive continental margins: Geology, v. 25, p. 631–634, doi: 10.1130/0091-7613(1997)025<0631:IORSPO>2.3.CO;2.

Weissel, J.K., and Seidl, M.A., 1998, Inland propagation of erosional escarpments and river profile evolution across the southeast Australian passive continental margin, *in* Tinkler, K.J., and Wohl, E.E., eds., Rivers over rock: Fluvial processes in bedrock channels: Washington, D.C., American Geophysical Union, Geophysical Monograph 107, 323 p.

Wende, R., 1999, Boulder bedforms in jointed-bedrock channels, *in* Miller, A.J., and Gupta, A., eds., Varieties of fluvial form: Chichester, Wiley and Sons, 521 p.

Whipple, K.X., Hancock, G.S., and Anderson, R.S., 2000a, River incision into bedrock: Mechanics and relative efficacy of plucking, abrasion and cavitation: Geological Society of America Bulletin, v. 112, p. 490–503, doi: 10.1130/0016-7606(2000)112<490:RIIBMA>2.3.CO;2.

Whipple, K.X., Snyder, N.P., and Dollenmayer, K., 2000b, Rates and processes of bedrock incision by the Upper Ukak River since the 1912 Novarupta ash flow in the Valley of Ten Thousand Smokes, Alaska: Geology, v. 28, p. 835–838, doi: 10.1130/0091-7613(2000)028<835:RAPOBI>2.3.CO;2.

Whiting, P.J., 1997, The effect of stage on flow and components of the local force balance: Earth Surface Processes and Landforms, v. 22, p. 517–530, doi: 10.1002/(SICI)1096-9837(199706)22:63.0.CO;2-M.

Whitney, M.I., 1983, Eolian features shaped by aerodynamic and vorticity processes, *in* Brookfield, M.E., and Ahlbrandt, T.S., eds., Eolian sediments and processes: Amsterdam, Elsevier Developments in Sedimentology, v. 38, p. 223–245.

Whittaker, J.G., and Jaeggi, N.M.R., 1982, Origin of step-pool systems in mountain streams: American Society of Civil Engineers, Journal of the Hydraulics Division, v. 108, p. 758–773.

Wigley, T.M.L., 1972, Analysis of scallop patterns by simulation under controlled conditions: A discussion: Journal of Geology, v. 80, p. 121–222.

Wohl, E.E., 1992a, Gradient irregularity in the Herbert Gorge of northeastern Australia: Earth Surface Processes and Landforms, v. 17, p. 69–84.

Wohl, E.E., 1992b, Bedrock benches and boulder bars: Floods in the Burdekin Gorge of Australia: Geological Society of America Bulletin, v. 104, p. 770–778, doi: 10.1130/0016-7606(1992)104<0770:BBABBF>2.3.CO;2.

Wohl, E.E., 1993, Bedrock channel incision along Piccaninny Creek, Australia: Journal of Geology, v. 101, p. 749–761.

Wohl, E.E., 1998, Bedrock channel morphology in relation to erosional processes, *in* Tinkler, K.J., and Wohl, E.E., eds., Rivers over rock: Fluvial processes in bedrock channels: Washington, D.C., American Geophysical Union, Geophysical Monograph 107, 323 p.

Wohl, E.E., 1999, Incised bedrock channels, *in* Darby, S.E., and Simon, A., eds., Incised river channels: Chichester, UK, Wiley and Sons, p. 187–218.

Wohl, E.E., 2000, Substrate influences on step-pool sequences in the Christopher Creek drainage, Arizona: Journal of Geology, v. 108, p. 121–129.

Wohl, E.E., and Achyuthan, H., 2002, Substrate influences on incised-channel morphology: Journal of Geology, v. 110, p. 115–120, doi: 10.1086/324207.

Wohl, E.E., and Grodek, T., 1994, Channel bed-steps along Nahal-Yael, Negev Desert, Israel: Geomorphology, v. 9, p. 117–126.

Wohl, E.E., and Ikeda, H., 1997, Experimental simulation of channel incision into a cohesive substrate at varying gradients: Geology, v. 25, p. 295–298, doi: 10.1130/0091-7613(1997)025<0295:ESOCII>2.3.CO;2.

Wohl, E.E., and Ikeda, H., 1998, Patterns of bedrock channel erosion on the Boso Peninsula, Japan: Journal of Geology, v. 106, p. 331–345.

Wohl, E.E., and Merritt, D.M., 2001, Bedrock channel morphology: Geological Society of America Bulletin, v. 113, p. 1205–1212, doi: 10.1130/0016-7606(2001)113<1205:BCM>2.0.CO;2.

Wohl, E.E., Greenbaum, N., Schick, A.P., and Baker, V.R., 1994, Controls on bedrock channel incision along Nahal-Paran, Israel: Earth Surface Processes and Landforms, v. 19, p. 1–13.

Wohl, E.E., Thompson, D.M., and Miller, A.J., 1999, Canyons with undulating walls: Geological Society of America Bulletin, v. 111, p. 949–959.

Zen, E.-a., and Prestegaard, K.L., 1994, Possible hydraulic significance of two kinds of potholes: Examples from the paleo-Potomac River: Geology, v. 22, p. 47–50, doi: 10.1130/0091-7613(1994)022<0047:PHSOTK>2.3.CO;2.

Manuscript Accepted by the Society 1 March 2005

Printed in the USA